THE END OF FREE WILL

How Technology Rewrites Human Choice

Table of Contents

Introduction

Free will has always been one of humanity's most treasured beliefs. It is the invisible foundation behind responsibility, ambition, morality, love, faith, and identity. We build our lives around the assumption that we are the authors of our decisions—that our choices come from within, shaped by reason, emotion, experience, and values. Even when we make mistakes, we still believe they were *our* mistakes. Even when life is unfair, we hold onto the idea that we can respond freely, change direction, and choose a different path. Free will is not just a philosophical concept. It is the psychological engine that makes human life feel meaningful.

But what happens when the forces shaping our decisions are no longer limited to family, culture, education, and personal struggle? What happens when the strongest influence on human behavior becomes a system that is engineered, invisible, personalized, and always active? A system that watches what you click, what you pause on, what you search, what you fear, what you desire, and what you avoid—then adjusts your environment accordingly. Modern technology has not simply entered our lives as a neutral tool. It has evolved into a powerful architecture of behavioral influence, quietly rewriting the conditions in which human choice is formed. The world we live in today is not just digital. It is *designed*—and that design increasingly decides for us.

The End of Free Will: How Technology Rewrites Human Choice is not a book about blaming technology for every human problem. It is not a rejection of progress, innovation, or the modern world. It is a serious exploration of how digital systems are reshaping the internal mechanics of decision-making. Most people assume their phone is simply a device. Their apps are simply services. Their feeds are simply entertainment or information. Their online experience feels like something they control. But beneath that surface is a deeper structure—an ecosystem built around one central goal: influence. Technology today is designed not just to serve your needs, but to capture your attention, predict your behavior, and guide your responses. In other words, it does not just *react* to your choices. It increasingly *creates* them.

A century ago, human influence was largely physical and local. Your environment was your neighborhood, your family, your workplace, your community, and your nation. Today, your environment follows you everywhere, through a screen that stays in your pocket. And that environment is not passive. It is intelligent. It adapts. It learns. It tests you. It rewards certain behaviors and discourages others, often without you even realizing it. Whether it is what you buy, who you date, what you believe, what you fear, what you trust, what you hate, or what you dream about—your digital world is continuously shaping your mental world.

This is not done through obvious force. In fact, the most effective control is not the control that feels like control. The most powerful manipulation is the one that feels like freedom. Technology does not usually threaten you into obedience; it seduces you into patterns. It does not take away your choices completely; it overwhelms you with options, then quietly pushes you toward the option that benefits the platform, the advertiser, the algorithm, or the system behind it. In the modern age, the loss of autonomy does not arrive like a prison door closing. It arrives like convenience, personalization, entertainment, and comfort—until one day you realize that your mind is reacting more than it is thinking.

To understand what is happening, we must first understand a painful truth: human free will has always been limited. The brain was never a perfectly rational machine. We are emotional creatures. We are habitual creatures. We are social creatures. We respond to reward and punishment. We are shaped by identity, trauma, desire, fear, validation, and survival instincts. Much of what we call "choice" is influenced by subconscious programming built from childhood, culture, and experience. This reality doesn't mean free will is an illusion, but it does mean it is fragile. Human choice is real, but it exists inside conditions—and if those conditions are remind-shaped intentionally, choice becomes easier to hijack.

Technology exploits this fragility. It takes advantage of the brain's shortcuts and weaknesses, then amplifies them at scale. It uses the psychology of attention, dopamine, social validation, fear, and repetition to create behavioral patterns that are profitable and predictable. The more predictable you are, the easier you are to monetize. The easier you are to monetize, the more the system invests in influencing you. The result is a world where your emotions are triggered strategically, your desires are curated, and your beliefs are reinforced through engineered reality loops. This is not conspiracy theory. This is business logic mixed with behavioral science. And it is happening so smoothly that most people never pause to question it.

Consider something as simple as scrolling. Many people believe they scroll because they are bored. They scroll because they want news. They scroll because they want entertainment. But the deeper truth is that scrolling is often not a decision—it is a reflex. It is a trained behavior. Platforms are built to remove stopping points, remove mental rest, and create infinite continuation. The brain is not designed for infinite stimulation, yet we are now surrounded by it. Every scroll trains us to seek novelty, to avoid silence, and to replace thinking with consumption. Over time, we lose more than time. We lose patience. We lose attention control. We lose internal silence. And without internal silence, free will cannot breathe.

Now add personalization to this environment. The internet used to be a place you explored. Today, it is a place that explores you. Algorithms do not just show you content—they study you through content. They measure what excites you, what angers you, what scares you, and what keeps you engaged. Then they feed you more of it. This is how the digital world becomes a psychological mirror. But it is not a mirror that reflects you truthfully. It is a mirror designed to magnify your strongest reactions. In this mirror, you are not shown what is most meaningful—you are shown what is most addicting. Over time, people become trapped in a version of reality that is tailored to their emotional triggers. This is how free will begins to shrink without anyone noticing.

As artificial intelligence grows more advanced, the problem intensifies. AI systems do not merely predict what you like—they predict what you will do. They forecast your behavior based on patterns you don't even know you have. They identify your personality traits, vulnerabilities, and preferences faster than you can name them yourself. In a world where machines know your habits better than your closest friends, the question is not whether free will is being challenged. The question is whether human choice can survive in an environment that is designed to steer it.

And then there is surveillance—one of the most under-discussed forces shaping modern behavior. Surveillance is not just about cameras and government monitoring. It is the culture of being tracked, recorded, analyzed, and judged. When people know they are being watched—directly or indirectly—they begin to change. They censor themselves. They obey norms more quickly. They avoid unpopular opinions. They choose safety over authenticity. Surveillance does not need to arrest you to control you; it only needs to make you *afraid* of consequences. In this atmosphere, freedom becomes cautious. Free will becomes filtered. And eventually, people stop asking what they truly want—they start asking what is permitted.

The core argument of this book is simple but unsettling: modern technology does not eliminate free will overnight. It redesigns the environment of choice until free will becomes harder to access. It overwhelms the mind, distracts the mind, pressures the mind, and programs the mind. It turns humans into reactive creatures—scrolling, clicking, consuming, seeking validation, and repeating patterns—while believing they are still choosing freely. This is not because humans are weak. It is because the system is strong.

However, this book is not written to create hopelessness. Awareness is power. Once you understand how your decisions are shaped, you can begin to reclaim control. You can rebuild attention. You can protect your mind from manipulation. You can design boundaries that restore your autonomy. You can think intentionally instead of reacting emotionally. The goal is not to escape technology completely, but to stop being unconsciously controlled by it. Technology should be a tool you use—not a system that uses you.

In the chapters ahead, we will explore how human decision-making actually works, why the brain is vulnerable to digital influence, and how algorithms and attention-based systems reshape behavior. We will dive into social media identity formation, persuasion engineering, AI-driven prediction, surveillance culture, and the collapse of privacy. Most importantly, we will uncover the deeper truth behind all of this: that free will is not just an abstract debate—it is a survival skill. In an age where environments are built to influence your mind, choosing freely becomes an act of resistance.

The end of free will is not a moment in the future. It is a process in the present. The question is not whether technology is changing human choice. It already is. The real question is whether you will notice it in time—and whether you will be the designer of your life, or the product of a system built to design you.

Chapter 1

The Illusion of Choice in the Modern Age

Chapter 1 begins with a reality that most people rarely question: **we live in a world full of choices, yet we may be less free than ever.** Modern life appears to celebrate freedom through endless options—what to wear, what to watch, what to buy, what opinions to adopt, and even how to present ourselves. But choice is not the same as free will. Having many options can create the *illusion* of control while the deeper forces shaping our decisions remain hidden. This chapter explores how freedom in the modern age is often defined by surface-level selection rather than conscious, independent decision-making.

To understand the illusion, we must first clarify **what free will really means**. Free will is not just picking between two alternatives; it is the ability to choose with awareness, guided by personal values and genuine intention rather than impulse, pressure, or manipulation. Yet human decision-making is rarely pure. It is influenced by emotions, habits, social expectations, and subconscious programming built over years. In the digital era, these influences are no longer accidental—they are engineered and amplified through attention-driven platforms, persuasive design, and constant stimulation. As a result, many decisions happen automatically, and people often mistake reaction for choice.

This chapter also highlights **the gap between feeling free and being free**. You may feel independent because no one is physically forcing you. But freedom can be reduced without chains. When your attention is constantly pulled, your beliefs are shaped by algorithms, and your preferences are guided by curated environments, the mind becomes easier to direct. The most effective influence is the kind that feels like your own idea.

Finally, Chapter 1 explains **why this debate matters more than ever**. Because technology is no longer just a tool—it is becoming an invisible system that shapes behavior at scale. If we do not understand how choice is being redesigned, we risk losing our autonomy without noticing. In a world that profits from predictability, protecting free will becomes a form of modern survival.

WHAT FREE WILL REALLY MEANS

Free will is a concept that almost everyone claims to understand, yet very few people ever take the time to define it clearly. We use the phrase casually in everyday life: "I chose this," "It was my decision," "I did what I wanted." But when we look deeper, choice is not as straightforward as it appears. Real free will is not simply the presence of options, nor is it the act of selecting one option over another. Free will is the ability to make decisions that genuinely arise from one's own awareness, values, and intentions, without being subtly pushed, emotionally manipulated, or unconsciously guided by forces outside of conscious control.

At the heart of free will lies a simple question: *Where do your decisions come from?* If they come from clarity and intention, then they are your own. If they come from impulses, habits, fears, or external pressures, then the decision may feel voluntary, but its origin is not fully yours. This distinction is crucial because most of what humans consider "choice" happens on autopilot. The human brain is a machine built for efficiency. It relies on shortcuts—habits, routines, emotional triggers, learned responses—to avoid the exhausting task of deep analysis every time a decision needs to be made. This system helps us survive, but it also means that a surprising amount of behavior is automatic rather than consciously chosen.

Free will, then, is not a default human setting. It is a capacity that must be protected and actively exercised. To choose freely, a person must be able to notice what is influencing them. They must recognize the pressures acting upon their attention, desires, fears, and emotions. Without that awareness, a decision may feel voluntary while being heavily shaped by external factors. This is especially true in the modern world, where influence is no longer subtle or occasional. It is constant, engineered, and personalized.

Historically, free will was challenged by social expectations, cultural norms, religion, or family structures. But today, the strongest forces shaping choice come from technology. Every digital platform—whether it is a social network, streaming service, search engine, or e-commerce site—is designed with one objective: influence. Not necessarily in a malicious sense, but certainly in a strategic sense. Digital systems are built to capture attention, predict behavior, and guide engagement. Their goal is not to help individuals think clearly but to keep them scrolling, clicking, reacting, and returning.

In such an environment, free will becomes more fragile. When algorithms know your habits better than you do, when your preferences are shaped by what is shown to you, and when your emotions are triggered by design, it becomes difficult to claim that your choices are entirely your own. The architecture of the digital world is built to nudge, steer, and sometimes even manufacture desire. The more predictable your behavior becomes, the easier it is for systems to guide it. The more guided it becomes, the less room is left for conscious intention.

But free will is not only threatened by external manipulation. It is also weakened internally. Humans often confuse comfort with freedom. When an action feels easy, familiar, or pleasurable, they assume it was chosen freely. But many comfortable behaviors are simply well-reinforced habits. Technology exploits this by removing friction. Autoplay removes the need to decide. Infinite scroll eliminates natural stopping points. Notifications decide what deserves attention. Over time, the absence of friction leads to the absence of reflection—and without reflection, free will cannot function.

To understand free will fully, we must acknowledge that simply *feeling* in control does not guarantee that we *are* in control. True freedom involves questioning the origin of our impulses, observing the forces shaping our attention, and recognizing the difference between internal desires and externally induced ones. Free will requires the ability to interrupt automatic behavior long enough to choose intentionally rather than reactively.

In its purest form, free will means becoming the author of your own decisions—not the product of your conditioning, not the outcome of your environment, and certainly not the predictable endpoint of an algorithm designed to influence you. It means stepping out of autopilot, noticing the pressures acting on your mind, and reclaiming the space between stimulus and response. That space is where free will lives, and in the modern world, that space is shrinking.

Understanding this is the first step to protecting it.

THE GAP BETWEEN FEELING FREE AND BEING FREE

One of the most misleading aspects of modern life is how easy it is to *feel* free while not actually *being* free. Freedom, for most people, is defined by the absence of visible force. If no one is physically stopping you, if there are no chains, no threats, no locked doors, then the assumption is that your choices are entirely your own. But in reality, true freedom is not about the absence of external control; it is about the presence of internal autonomy. And internal autonomy can be quietly weakened without a person ever noticing.

This gap between feeling free and being free is most evident in how people interpret their daily decisions. When someone picks up their phone, scrolls for an hour, chooses what to watch, what to buy, or what to believe, they often assume these decisions are freely made. They feel spontaneous, voluntary, and self-driven. Yet behind these actions lies a complex system of psychological triggers, habit loops, emotional cues, and algorithmic influences that shape the decision before it even reaches conscious awareness. The person feels free because the influence is invisible.

Human beings have always been vulnerable to subtle forms of influence, but the digital world magnifies this vulnerability. Modern platforms are built around interfaces designed to capture attention, stimulate emotion, and reinforce behavior. They know when you are tired, when you are bored, when you are anxious, and when you are most likely to react. They exploit predictability, not through force, but through personalization. When your environment adapts to your weaknesses, your impulses can be mistaken for genuine desires.

This creates a paradox: the more the world caters to your preferences, the less those preferences remain your own.

Feeling free, therefore, is often the result of comfort. You feel free because the digital world is frictionless. With a swipe, a tap, or a click, you can access entertainment, information, communication, or distraction. Convenience becomes a kind of psychological anesthesia. It numbs awareness. It creates a sense of ease that masks the deeper truth: choices are being shaped long before you make them. The illusion of freedom grows stronger as actual autonomy grows weaker.

Being free, by contrast, requires the ability to pause. It requires the mental space to examine what is influencing you. A person who is truly free can recognize when they are acting out of habit rather than intention, when emotions are steering their behavior, and when external systems are shaping their desires. True freedom requires the courage to observe one's own impulses without obeying them automatically.

The gap between feeling free and being free becomes especially visible when we consider identity. People often believe that their preferences—what they like, what they dislike, what they support, what they oppose—reflect their authentic selves. But in a world where algorithms curate reality, people are increasingly shaped by the content they consume. Each scroll reinforces a belief. Each click strengthens a preference. Over time, a person may become deeply attached to attitudes that were never chosen consciously. They feel free because their identity feels familiar, but familiarity is not the same as authenticity.

Another layer of this gap lies in social behavior. Online worlds encourage performance. People shape their actions around visibility, approval, and social validation. They say what is safe, share what is expected, and avoid what may provoke backlash. They feel free because they willingly participate, but participation is often driven by fear of judgment or desire for belonging. A mind that constantly seeks approval is not a free mind; it is a mind negotiating its own survival.

Perhaps the greatest danger is that this loss of freedom does not shock or alarm. It happens gradually, quietly, without confrontation. The systems shaping behavior do not restrict choice—they redirect it. They do not limit options—they present the most profitable ones. They do not silence people—they overwhelm them with noise. In such an environment, the absence of force creates a false sense of freedom.

Being free requires awareness of these dynamics. It requires questioning the origin of your desires, examining your reactions, and resisting the seduction of convenience. Feeling free is effortless; being free is a discipline. Feeling free requires no awareness; being free requires constant awareness. Feeling free is passive; being free is active.

The modern world has mastered the art of making people feel free while quietly shaping their choices. The challenge of our time is to bridge that gap—to transform the illusion of freedom into the reality of autonomy.

WHY THIS DEBATE MATTERS MORE THAN EVER

There have always been debates about free will—philosophers, theologians, and scientists have argued for centuries about whether humans truly control their own actions. But never in history has the question been as urgent as it is today. The reason is simple: for the first time, the environment shaping human decision-making is not natural, cultural, or spiritual. It is engineered. It is deliberate. It is intelligent. And it is evolving faster than human awareness can keep up with. The debate about free will is no longer abstract or academic. It has become a question about the future of human autonomy in a world built to influence.

In earlier eras, human choices were shaped by family, society, religion, and basic survival needs. These forces were powerful, but they were understandable. People knew when they were being pressured, judged, or guided. Today's influences are different. They operate in silence. They analyze behavior at scale. They anticipate reactions before they occur. They create environments so personalized that individuals cannot easily distinguish their own desires from the desires manufactured around them. This shift is why the debate matters more deeply than it ever has: because influence has become invisible, and invisible influence is the most difficult to resist.

Another reason this debate is critical today is the role of data. Every digital trace—every click, pause, purchase, comment, hesitation, and search—is recorded, analyzed, and used to refine models that predict human behavior. These predictions do not remain predictions for long; they become tools of persuasion. When a system knows what you want before you consciously know it, the line between choice and manipulation becomes blurry. And this blurriness has consequences far beyond entertainment or consumer behavior. It shapes political beliefs, moral judgments, emotional states, and even perceptions of reality.

We are living in a moment where personalized realities are replacing shared ones. Algorithms decide what news people see, what stories gain attention, what truths gain traction, and what narratives fade into silence. Two people standing in the same room can inhabit completely different informational worlds, each shaped by their digital history. In such a fragmented landscape, the ability to think independently becomes harder, not easier. The debate about free will is therefore tied to the health of society itself. A population that cannot think freely cannot act freely, and a population that cannot act freely becomes easy to govern, easy to divide, and easy to manipulate.

This debate matters today because technology does not merely influence behavior—it speeds it up. The pace of life has increased to the point where reflection often feels unnatural. People consume information faster than they can process it, react before they can understand, and form opinions before they have examined them. In this accelerated environment, free will is not lost to force; it is lost to momentum. The world moves quickly, and when thinking slows down, external systems fill the space with ready-made conclusions.

Another reason this discussion is urgent is that the tools influencing human behavior are becoming more intelligent by the day. Artificial intelligence does not just learn from human patterns; it optimizes them. It creates loops that reinforce behavior, reward impulsivity, and strengthen emotional responses. It identifies vulnerabilities that even individuals may not recognize—loneliness, insecurity, boredom, anxiety—and uses those emotional states to increase engagement. This is not a conspiracy; it is an economic strategy. But economic strategies have psychological consequences, and psychological consequences have philosophical implications. If technology learns how to shape desire, then desire itself becomes a product.

Yet perhaps the most important reason this debate matters now is that people are unaware of what is at stake. Most individuals understand privacy risks or screen-time addiction on a surface level, but they do not grasp the deeper shift occurring beneath it. What is being redefined today is not simply behavior—it is human agency. The structures that once supported free will—attention, reflection, internal silence, self-awareness—are being eroded by constant noise, endless content, and digital overstimulation. When the mind is continually occupied, it becomes reactive. When it becomes reactive, it becomes programmable.

The urgency of this debate comes from recognizing that free will is not a guaranteed human attribute. It can weaken quietly. It can shrink. It can dissolve under the weight of convenience, distraction, and manipulation. And once lost, it is difficult to reclaim because individuals no longer recognize what it feels like to choose freely.

In a world where technology rewrites human choice, the debate about free will becomes a debate about human dignity. It becomes a question about whether the future will belong to autonomous individuals or to systems that understand humans better than humans understand themselves. That is why this discussion matters—not tomorrow, not someday, but now.

Chapter 2

The Science Behind Human Decisions

Human decisions often feel personal, intentional, and deliberate. When people choose a career, respond to a message, form an opinion, or pick a meal, they usually believe the decision came from a place of conscious control. Yet modern neuroscience paints a very different—and far more complex—picture of how the mind actually works. Beneath every decision lies a sophisticated network of emotional responses, habitual patterns, predictive mechanisms, and subconscious signals that shape human behavior long before awareness enters the scene. In other words, the brain does not simply *make* decisions—it prepares them, scripts them, and often executes them automatically.

This chapter explores the scientific foundations behind human choice: how the brain functions as a decision-making machine, how emotions and habits silently guide behavior, and why most daily choices happen without conscious thought. Instead of imagining ourselves as rational agents carefully analyzing every situation, science reveals that we operate much more like pattern-driven organisms responding to cues, triggers, and hidden influences embedded in both our biology and environment.

The goal of this chapter is not to diminish human intelligence, but to illuminate the invisible forces shaping every choice we make. When we understand the mechanisms behind decision-making, we gain the ability to recognize when we are acting automatically and when we are truly choosing. This awareness becomes especially critical in today's world, where technology, advertising, social pressure, and digital design strategically tap into the same systems that evolved for survival.

To understand how technology rewrites human choice, we must first understand how the brain writes its own decisions.

THE BRAIN AS A DECISION MACHINE

The human brain is often romanticized as a center of creativity, imagination, and meaning, but beneath all of its complexity, it is fundamentally a decision machine—an organ designed to evaluate information, predict outcomes, and choose actions that support survival. Every moment, whether you are aware of it or not, the brain is analyzing the world, estimating risks, weighing probabilities, and selecting responses. This process is constant and relentless, occurring even when you are not consciously focused on making a choice. Understanding this machinery is essential because it reveals something profound: choice does not emerge from a single moment of intention. It emerges from a continuous chain of mental events that happen both inside and outside conscious awareness.

At its core, the brain is built to conserve energy. Thinking deeply is expensive, metabolically speaking. Reflection, reasoning, and analysis require significant effort, so the brain prefers shortcuts whenever possible. These shortcuts operate through neural pathways that grow stronger with repetition. In other words, the brain becomes more efficient by turning repeated decisions into automatic patterns. Once a pattern is formed, the brain no longer evaluates it each time; it simply activates the established circuit. This is useful for survival—quick reactions help humans avoid danger, find food, and navigate a complex world without being overwhelmed by constant deliberation.

This design reveals why the brain functions as a prediction engine. Before making any decision, the brain compares the current situation to past experiences. It looks for familiar cues: a tone of voice, a type of environment, a certain facial expression, a remembered feeling. Based on these cues, it predicts the likely outcome of different actions. These predictions happen in milliseconds, far too quickly for conscious awareness to catch up. By the time you "decide," the brain has already calculated which response seems safest or most rewarding based on past learning.

The decision-making process can be divided into two broad systems. The first is fast, automatic, and emotional. It handles most of the everyday choices humans make—when to speak, how to interpret a gesture, when to pick up the phone, how to respond to a message. This system is driven by instinct, memory, and learned associations. The second system is slow, deliberate, and rational. It is responsible for long-term planning, self-reflection, and conscious analysis. It is the part of the brain we like to believe is always in charge, but in reality, it intervenes far less often than we imagine.

Most of the time, the rational system only steps in when something interrupts the automatic flow—when something unexpected happens, when a problem arises, or when a decision carries significant consequences. Yet even then, the slow system does not operate independently. It is influenced by emotional signals, bias patterns, and subconscious associations formed long before the conscious mind becomes involved. This is why even the most logical decisions often carry emotional undertones.

The brain's structure also reinforces its role as a decision machine. The prefrontal cortex, responsible for judgment and planning, works closely with the limbic system, which handles emotions, rewards, and instinctive reactions. These regions are constantly communicating, negotiating between impulse and intention. That negotiation creates what we experience as a "decision." But much of the negotiation happens before the conscious self becomes aware of it. By the time you register a desire, preference, or impulse, your brain has already run through countless evaluations.

Importantly, the brain does not choose in a vacuum. It is influenced by internal states—hunger, fatigue, stress, hormones—as well as by external conditions like environment, social pressure, and technology. Every notification, advertisement, sound, and image becomes a data point that the brain must interpret. In the digital age, where stimuli are constant and carefully engineered, the brain is forced to make far more micro-decisions than it ever evolved to handle. This overwhelms the rational system and strengthens the dominance of automatic processing.

Recognizing the brain as a decision machine does not diminish the value of human experience; it reveals the mechanisms behind it. When you understand how decisions form, you can better recognize which ones are truly yours and which ones are simply the result of conditioning, instinct, or external influence. Awareness of the brain's machinery does not eliminate its automatic tendencies, but it allows you to interrupt them—to insert conscious intention into a process that typically runs without your input.

In a world designed to capture attention and shape behavior, understanding the brain's decision architecture becomes not just important but essential. It allows you to reclaim the space where choice truly lives.

EMOTIONS, HABITS, AND HIDDEN DRIVERS

Human beings like to imagine that their choices are built on logic, reflection, and conscious thought. But beneath the surface of every decision lies a network of emotional signals, habitual patterns, and subconscious influences that quietly steer behavior long before awareness enters the picture. These forces do not announce themselves. They operate in the background, shaping preferences, reactions, and intentions in ways that often feel natural but are far from neutral. To understand how much of life is driven by these hidden systems, we must examine how emotions and habits work together to form the foundation of human decision-making.

Emotions are often misunderstood as fleeting feelings or irrational impulses, but in reality, they are survival mechanisms. They evolved to help humans detect danger, recognize opportunity, form social bonds, and respond quickly to complex situations. Long before the rational mind can evaluate a situation, the emotional brain has already determined whether something feels safe, threatening, rewarding, or uncertain. These emotional evaluations influence perception itself. When something feels familiar or comforting, the brain labels it as valuable. When something feels unfamiliar or overwhelming, the brain labels it as risky. These emotional tags shape every decision that follows, yet they often occur outside conscious awareness.

For example, a person may "choose" a job, a partner, or a belief system not because it is objectively the best option but because it aligns with an emotional memory—something that feels familiar, safe, or validating. Humans rarely question these emotional signatures because they appear as intuition or preference. But intuition is often just the brain's emotional shorthand for past experiences. The more emotionally charged an experience is, the more powerfully it shapes future decisions. Trauma, for instance, leaves deep emotional imprints that influence behavior long after the threat is gone. Comforting experiences do the same—they become psychological magnets that pull a person toward similar situations.

Habits form another layer of hidden influence. A habit is not just a repeated action; it is a learned pattern that the brain executes automatically to conserve energy. Once a habit loop is established, it becomes the default response to certain cues. This means you may believe you are choosing to act in a particular way when, in reality, you are simply repeating a pattern created months or years earlier. Habits form through three components: the cue, the routine, and the reward. The cue triggers the behavior, the routine carries it out, and the reward reinforces the loop, making the brain more likely to repeat it. Over time, the brain stops questioning whether the behavior is helpful. It simply follows the path of least resistance.

These habits intertwine with emotions in subtle ways. Many habitual behaviors begin as emotional responses—eating to soothe stress, scrolling to escape boredom, procrastinating to avoid fear of failure. The emotional relief becomes the reward that reinforces the behavior. Eventually, the brain connects the emotional cue to the habitual action so strongly that the behavior occurs automatically. This is why people often engage in behaviors that contradict their long-term goals. The habit is not serving the conscious mind; it is serving the emotional mind.

Beyond emotions and habits, there are deeper subconscious drivers that influence behavior. These include identity beliefs ("I am this kind of person"), internal narratives ("This is how life works"), and early conditioning from childhood. These hidden drivers form the lens through which individuals interpret the world. If someone believes they are unworthy, they may avoid opportunities subconsciously. If someone believes the world is dangerous, they may make conservative choices even when risk is minimal. These beliefs create patterns that feel like personality traits but are often just unexamined assumptions shaped by experience.

Another hidden driver is social conditioning. Humans are deeply influenced by the expectations of those around them. Conforming is often subconscious because belonging is tied to emotional survival. People absorb norms without questioning them—what success looks like, what relationships should be, what opinions are acceptable. These norms shape decisions so naturally that individuals rarely notice the pressure.

The modern world amplifies these hidden influences. Digital platforms exploit emotional responses, reinforce habits through constant cues, and shape subconscious drivers through repeated exposure. When a person encounters the same message multiple times online, they are more likely to believe it—not because it is true, but because repetition makes information feel familiar. This is how opinions form without conscious thought.

Understanding emotions, habits, and hidden drivers does not eliminate their power, but it allows individuals to recognize when these forces are controlling behavior. Awareness transforms automatic reactions into opportunities for intentional action. The goal is not to suppress emotions or eliminate habits, but to understand them deeply enough that they no longer operate in secret.

WHY MOST CHOICES ARE AUTOMATIC

Most people believe they think through their decisions carefully. They imagine themselves weighing options, considering consequences, and choosing the path that aligns with their desires and goals. But the truth is far more complex—and in many ways, more unsettling. The vast majority of human choices are not the result of deliberate thought. They are automatic responses shaped by patterns, cues, emotions, memories, habits, and environmental triggers that the brain processes long before the conscious mind becomes aware. The feeling of intentional decision-making often comes after the brain has already acted.

The automatic nature of most choices begins with the way the brain is structured. The brain must process massive amounts of information every second—far more than we could consciously manage. To keep us functioning, it delegates the majority of tasks to systems that operate beneath awareness. These systems learn from repetition, adapt to routines, and create shortcuts to save energy. As a result, the brain handles everyday life through autopilot modes. You do not choose to breathe, blink, walk a certain way, or react instinctively to sudden movement. But this automaticity extends far beyond basic functions; it includes behaviors most people assume are conscious.

For instance, consider how people navigate their routines. The route they take to work, the apps they check first in the morning, the tone they use when speaking to certain people, the way they respond to stress—all of these are shaped by repeated patterns. Once a behavior becomes familiar, the brain no longer evaluates it. It simply executes the script. This is why so many decisions feel effortless—not because they require no thought, but because the thinking happened long ago, during the formation of the habit.

Another reason choices are automatic is that the brain reacts faster than conscious awareness can keep up. Studies in neuroscience reveal that decisions can be detected in brain activity seconds before a person believes they made a choice. The conscious mind often steps in only after the action has already been set in motion, creating a narrative that makes the decision feel intentional. This storytelling function helps maintain a sense of identity and coherence, but it does not accurately reflect the true sequence of events.

Environmental cues also play a powerful role in automatic decisions. A notification sound, a familiar scent, a certain posture, or even the lighting in a room can trigger behaviors without conscious input. Restaurants use lighting and music to influence eating pace. Stores use layout to guide movement. Apps use sound and color to capture attention. People rarely notice these cues, yet they respond to them instinctively. A person may pick up their phone because they felt a phantom vibration, not because they wanted to check anything. They may open the fridge because they walked into the kitchen, not because they were hungry. These actions feel natural, but they are triggered, not chosen.

Emotion further accelerates automatic decision-making. When a person is stressed, angry, afraid, or excited, the emotional brain takes control. It prioritizes speed over accuracy, leading to quick, instinctive choices. This is useful in dangerous situations, but in everyday life, it means many decisions are made while the rational mind is sidelined. A person may snap at someone, make an impulsive purchase, or avoid a challenging task—not because they thought it through, but because emotion dictated the response.

Social influence also drives automatic behavior. Humans are wired for imitation. When people see a crowd doing something, they are more likely to follow. When they hear applause, they feel compelled to clap. When they notice a trend, they feel pressure to participate. Much of this happens without conscious intention. The desire to belong is one of the strongest forces shaping human behavior, and it often overrides personal choice.

Technology further pushes decisions into automatic mode. Infinite scroll, autoplay, curated feeds, and personalized recommendations remove the need for conscious selection. The next video plays before a person has time to think. The next item appears before they consider alternatives. The next post surfaces before they question why. These designs are not accidental—they train the brain to follow digital pathways instinctively. Over time, a person becomes less aware of why they click, watch, or consume. The decision feels voluntary, but the behavior is conditioned.

The reality is not that humans lack free will, but that free will is not the dominant mode of operation. Conscious choice is the exception, not the rule. It appears when something disrupts the automatic flow—when a habit breaks, when an emotion stabilizes, or when a person deliberately pauses to examine their impulses.

Understanding how much of life runs on autopilot is not meant to discourage. Instead, it provides a pathway to reclaiming agency. When you recognize that most choices are automatic, you can begin to interrupt them—to replace reactive behavior with intentional action. Awareness becomes the key to unlocking true freedom.

Chapter 3

How Technology Enters the Mind

Technology no longer sits outside the human experience as a neutral tool; it has become an active participant in shaping how people think, feel, and behave. Screens are not just devices we look at—they are environments we enter. They surround the mind with endless information, emotional triggers, social cues, and persuasive designs that slowly rewire the mechanics of decision-making. What makes modern technology so influential is not its speed or sophistication, but its intimacy. It meets the mind where it is most vulnerable: in moments of boredom, loneliness, stress, curiosity, and distraction.

This chapter explores the psychological architecture of digital life—the way screens create immersive mental worlds, how comfort-driven designs weaken self-regulation, and why attention has become the most valuable and contested resource of the 21st century. Technology does not force itself into the mind; it slips in quietly through convenience, entertainment, and personalization. It offers ease, instant gratification, connection, and stimulation—until the mind adapts to the point where silence feels uncomfortable and stillness feels unnatural.

We will examine how digital platforms transform everyday environments into psychological landscapes designed to guide emotion and behavior. We will also look at how the pursuit of convenience often results in the erosion of autonomy, as choices become shaped by algorithms rather than personal intention. Finally, we will uncover why attention—once a private internal resource—has become a global marketplace where companies compete ruthlessly for every second of mental presence.

Understanding how technology enters the mind is essential to understanding how it reshapes free will. Only by recognizing the invisible influence of screens, comfort, and attention-driven design can we begin to reclaim control over our mental environment.

SCREENS AS PSYCHOLOGICAL ENVIRONMENTS

Screens have become the primary gateway through which humans interact with the world, but they are far more than simple display devices. Every screen—whether it is a smartphone, tablet, laptop, or television—functions as a psychological environment, a space that shapes perception, behavior, emotion, and identity. When a person looks at a screen, they are not merely observing digital content; they are entering a mental landscape crafted to engage, influence, and hold attention. Over time, these environments become so familiar that individuals forget how profoundly they shape internal experience.

A physical environment influences how people feel and behave. We know that lighting affects mood, that room structure shapes movement, and that physical noise affects concentration. What most people fail to recognize is that screens, too, create environments—just not physical ones. They produce psychological spaces filled with symbols, colors, motion, faces, narratives, and emotional cues. Within seconds, these digital environments can shift a person's mental state. A notification induces anticipation. A viral video sparks laughter. A disturbing news clip produces anxiety. The screen becomes a portal into emotional worlds that feel real because they activate the same neural circuits as real-life experiences.

Unlike physical spaces, digital environments adapt continuously. They are curated specifically for each user, learning their preferences, behaviors, fears, desires, and weaknesses. This makes screens psychologically immersive in ways that physical spaces can never be. A person walking into a room sees the same walls and furniture as everyone else. But a person opening their phone sees a world tailored to them—news chosen for them, ads crafted for them, entertainment shaped around them. The environment is not shared; it is personalized. This personalization creates a powerful illusion of relevance while subtly narrowing the range of thoughts and perspectives a person encounters.

Screens also dissolve the boundary between public and private space. A bedroom becomes a workplace. A kitchen becomes a marketplace. A living room becomes a social arena. The screen transforms every physical location into a psychological hybrid, blending mental states that would normally remain separate. This "environmental collapse" affects how people experience time, emotion, and focus. The brain evolved to shift between environments gradually, but screens force rapid transitions. In one minute, a person may move from humor to fear, from intimacy to outrage, from relaxation to stress—all within the same physical location. This constant environmental shifting exhausts cognitive resources and weakens emotional regulation.

Another defining trait of screens as psychological environments is their ability to compress or distort reality. Digital environments accelerate time, collapsing complex events into short clips and headlines. They amplify emotional extremes, rewarding content that shocks, angers, entertains, or provokes. They present idealized images of beauty, success, and happiness that shape self-perception. They create the illusion of connection through likes and messages while often deepening loneliness. In these environments, the mind internalizes unrealistic expectations and distorted norms because it encounters them repeatedly until they appear natural.

Screens also encourage passive absorption rather than active participation. The user becomes a spectator in a world designed by others, guided by algorithms, suggestions, and automated sequences. Autoplay, infinite scroll, and recommendation systems transform screens into psychological traps, keeping the mind suspended in a state of constant consumption. As a result, the brain begins associating screens with effortless stimulation. Over time, real-world experiences—conversations, work, silence, self-reflection—feel less engaging because they require more cognitive effort.

Importantly, digital environments shape attention itself. They are filled with visual cues engineered to capture focus: movement, color contrast, sudden sound, emotional faces, bold text. These cues hijack neural circuits that evolved for survival, making the screen feel urgent even when nothing meaningful is happening. The more time spent in these environments, the more the brain adapts to their demands. This adaptation changes attention span, reward sensitivity, and tolerance for boredom—fundamental aspects of the psychological self.

Finally, screens do not just reflect behavior; they train it. They condition the user through feedback loops—notifications reward return visits, likes reinforce posting, recommendations shape curiosity. Over time, the screen becomes not just a place where decisions happen but a place that generates the conditions leading to those decisions.

Screens are not passive tools. They are psychological habitats that reshape the mind. To understand how technology influences free will, one must first understand that entering a screen means entering a world built for influence.

DIGITAL COMFORT VS. MENTAL CONTROL

Comfort has always been a fundamental human desire. People naturally gravitate toward ease, security, and convenience because the brain is wired to conserve energy and avoid unnecessary strain. In the digital age, technology has transformed comfort from a luxury into a default expectation. With a few taps, one can order food, find entertainment, connect with others, or escape discomfort. Everything is designed to be effortless. On the surface, this feels like progress. But beneath this convenience lies a deeper psychological shift: the more we rely on digital comfort, the less control we retain over our own minds.

Digital comfort operates on a simple principle—reduce friction. Friction is anything that requires thought, patience, or effort. Technology removes friction at every turn. Autoplay removes the effort of choosing the next video. Infinite scrolling eliminates the need to stop and decide whether to continue. Notifications ensure the mind never fully rests. Suggestions replace independent searching. Algorithms curate choices so thoroughly that the need to think becomes optional. What begins as a convenience gradually becomes a dependency.

The danger of digital comfort is not that it makes life easier, but that it weakens the mental muscles required for self-regulation. When everything is instant, patience fades. When everything is personalized, curiosity narrows. When stimulation is always available, boredom becomes intolerable. These changes affect more than mood—they reshape decision-making. A mind accustomed to constant ease begins to resist anything that requires sustained attention, effort, or introspection. Tasks that once felt manageable now feel overwhelming simply because they lack the frictionless appeal of digital alternatives.

Technology companies understand this dynamic well. Comfort is profitable because it keeps people engaged. The easier something is, the more likely users are to continue doing it. Platforms are designed to anticipate needs before they arise and satisfy them before users have time to think. This creates a cycle of passive engagement. Instead of seeking meaning, people seek ease. Instead of making conscious choices, they follow whatever requires the least effort.

But comfort comes with a hidden cost: the surrender of mental control. Mental control is the ability to direct attention, regulate emotion, manage impulses, and make intentional decisions. It is built through practice—through moments of discomfort, reflection, and challenge. When digital comfort removes these moments, the mental processes that sustain autonomy weaken. People begin reacting instead of choosing, consuming instead of engaging, following instead of thinking.

One of the most significant impacts of digital comfort is its effect on emotional resilience. Discomfort is a natural part of human experience. Boredom, loneliness, frustration, uncertainty—these emotions once served as signals that helped people evaluate their lives and grow. Now, they are quickly numbed by digital distraction. The moment an uncomfortable feeling arises, a screen can drown it out. Over time, the mind loses tolerance for discomfort. It becomes easier to escape emotions than to process them. This avoidance disrupts personal growth and weakens the sense of self-agency.

Digital comfort also narrows attention. When the brain becomes trained to expect instant gratification, it struggles to maintain focus on tasks that require sustained effort. Reading becomes difficult. Deep thinking feels tiring. Creativity declines. Complex conversations become draining. In this environment, the ability to control one's own mind erodes while the influence of external systems grows stronger.

This imbalance becomes even more concerning when we consider how digital comfort intersects with persuasive design. When a user is relaxed, passive, and accustomed to ease, they are more susceptible to suggestions. Recommendations feel helpful. Advertisements feel personalized. Subtle nudges feel natural. The line between desire and manipulation becomes blurred because the mind is no longer actively evaluating choices; it is simply accepting what feels effortless.

Ultimately, digital comfort becomes a silent trap. It creates the illusion of control while quietly reducing it. It offers relief while weakening resilience. It promises freedom while reinforcing dependence. The more comfortable the digital world becomes, the more control individuals unknowingly hand over to the systems shaping their behavior.

Regaining mental control does not require rejecting technology. It requires recognizing that comfort, while pleasant, is not neutral. It changes how the mind functions. And if we do not actively balance comfort with intentional effort, our ability to choose freely will continue to fade.

ATTENTION AS THE NEW BATTLEFIELD

In every era of human history, power has centered around the ability to control something valuable—land, resources, labor, information. But in the digital age, the most valuable resource is no longer physical or even informational. It is human attention. Attention determines what people think about, what they care about, what they believe, and ultimately, who they become. This makes it the most contested territory of the 21st century, a battlefield where corporations, platforms, advertisers, governments, and influencers compete relentlessly for every second of mental presence. The struggle for attention is not metaphorical—it is psychological warfare fought through design, data, and emotion.

Attention is finite. The brain can only consciously focus on one thing at a time, and even that focus is fragile. Modern digital systems exploit this fragility. They are engineered to attract attention quickly, hold it as long as possible, and redirect it before it drifts elsewhere. Every color, vibration, sound, headline, and recommendation is crafted to penetrate the mind's defenses. Notifications interrupt thought. Autoplay interrupts intention. Alerts interrupt silence. The result is a constant assault on the cognitive space where free will should operate.

What makes this battlefield so dangerous is that the weapons used are deeply aligned with human psychology. The brain is wired to respond to novelty, movement, fear, reward, and social cues. Digital platforms amplify these triggers to pull attention away from anything slow, quiet, or demanding. A thoughtful conversation or a complex task cannot compete with the instant gratification provided by curated feeds and rapid-fire content. Attention becomes fragmented, scattered across hundreds of small digital demands. Over time, this fragmentation weakens the ability to sustain focus, making it even easier for external systems to control attention.

The battle for attention also shapes emotional experience. Outrage, fear, excitement, and shock capture attention more effectively than calm or neutrality. As a result, digital environments amplify extreme content because extreme emotions prolong engagement. People think they are choosing what to watch or read, but much of what appears on their screens is selected because it triggers emotions that hold attention tightly. Emotional manipulation becomes a strategic tool, not because platforms want people to feel a certain way, but because emotional intensity increases the time spent on the platform.

Another dimension of this battlefield is personalization. Algorithms learn what captures your attention, then deliver more of it. This creates an attention loop. The more a person engages with a topic, tone, or idea, the more similar content they receive. This narrows awareness, creating tunnel vision. Instead of exploring the world, individuals explore a customized slice of it—one that mirrors their existing interests and reactions. This does not just shape attention; it shapes identity. People feel drawn toward certain ideas not because they chose them consciously, but because those ideas were the ones most effective at holding their attention.

Attention has also become monetized. Every second spent looking at a screen generates data, and data is converted into profit. Platforms are rewarded for capturing and directing attention, not for supporting well-being or autonomy. This financial incentive accelerates the competition for attention, pushing platforms to design increasingly persuasive and addictive systems. The user is no longer the customer; the user is the product. Their attention is what is bought, sold, and traded.

This battlefield does not only affect individuals. It affects society. When attention becomes fragmented, collective focus dissolves. Important issues struggle to hold attention. Complex conversations are overshadowed by viral noise. Public discourse becomes polarized because emotionally charged content dominates. People no longer engage deeply with ideas; they skim them, react to them, or scroll past them. A society that cannot sustain attention cannot sustain meaningful thought.

Regaining control of attention is therefore essential to preserving autonomy. Attention is the gateway to everything the mind can do—think, question, reflect, remember, and decide. When attention is captured, these capacities weaken. When attention is reclaimed, they strengthen. The battle for attention is not fought through force but through awareness. It requires recognizing how digital environments pull us, how emotions guide our clicks, and how habits shape our engagement.

Attention is the foundation of free will. Without control over it, choice collapses. Understanding this battlefield is the first step toward reclaiming the power to direct one's own mind.

Chapter 4

Algorithms That Predict You

In the past, understanding human behavior required observation, intuition, and time. People revealed themselves slowly—through patterns, words, choices, and habits. Today, algorithms perform the same task in seconds. They do not rely on intuition or guesswork. They rely on data—millions of tiny signals collected from every click, pause, purchase, message, search, and hesitation. These signals form behavioral fingerprints so precise that algorithms can often predict what a person will do before they consciously know it themselves. In this new digital landscape, prediction is not merely a technological capability; it is a form of power.

This chapter explores how algorithms learn, anticipate, and ultimately shape human behavior. What makes modern prediction systems so influential is not their accuracy alone, but their ability to personalize every digital experience. Platforms no longer show the same content to everyone. Instead, they curate a unique reality for each individual—one that reflects their fears, desires, weaknesses, and emotional triggers. Over time, this personalization creates psychological feedback loops that influence how people think, feel, and act. The algorithm becomes a quiet architect of identity.

Yet the most important shift occurs when prediction begins to transform into direction. When an algorithm forecasts what you are likely to choose, it can also guide you toward that choice through subtle cues: suggestions, recommendations, notifications, or emotional prompts. Action becomes easier than reflection, and the predicted behavior becomes the chosen behavior.

Understanding algorithms is no longer optional; it is necessary for preserving autonomy. This chapter reveals how predictive systems work, how they reshape identity, and why the moment an algorithm can predict you is also the moment it begins to influence you.

DATA PATTERNS AND BEHAVIORAL FORECASTING

Human behavior has always followed patterns. Even before technology existed, people showed predictable tendencies—habits formed from repetition, emotional responses shaped by past experience, preferences rooted in personality and culture. But identifying these patterns required careful observation, intuition, and time. Today, algorithms perform this task with astonishing speed and precision, detecting behavioral patterns at a scale the human mind could never achieve. This is the foundation of behavioral forecasting: the ability of digital systems to predict what a person will do, want, believe, or buy before they consciously recognize it themselves.

The process begins with data—seemingly insignificant traces of daily behavior. Every search, swipe, pause, like, purchase, and message becomes a datapoint. Individually, these datapoints may appear meaningless, but collectively, they form a behavioral mosaic. Patterns emerge not from single actions but from thousands of micro-actions. An algorithm does not care why someone hesitates before clicking; it only cares that they hesitated. That hesitation becomes part of a psychological profile. Over time, the system learns what hesitation means, when it appears, and what typically follows it.

This is where algorithms excel: they detect relationships between behaviors that no human observer could ever notice. For example, they may learn that people who watch certain videos late at night are more impulsive buyers the next morning. Or that someone who repeatedly pauses on content about self-improvement is also likely to engage with motivational products. Or that a slight change in scrolling speed often signals emotional distress or boredom. These patterns emerge naturally from data—algorithms simply identify them, measure them, and use them to predict the future.

Behavioral forecasting becomes even more powerful when predictive models begin learning from similar users. Algorithms do not just analyze one person in isolation; they compare them to millions of others whose data exhibits parallel tendencies. This allows them to predict not only what a person might do next, but what they might become. If someone begins showing early signs that match the trajectory of hundreds of other individuals, the algorithm can anticipate future behavior even before it appears in the person's own actions.

Importantly, forecasting does not require deep personal information. People worry about privacy in terms of names, addresses, or messages. But prediction thrives on patterns, not identities. An algorithm does not need to know a person's private secrets; it only needs to know how they behave. This shift—from knowing who you are to knowing what you will do—marks a profound transformation in the role of technology in human life.

Behavioral forecasting also influences the digital environment itself. Predictions guide what content appears on a feed, what ads are shown, what recommendations surface, and what messages are prioritized. This creates a feedback loop. The system predicts what will capture attention, shows it, gathers more data from engagement, strengthens the prediction, and then shows even more precise content. The user believes they are freely interacting with the digital world, but the world is increasingly shaped by predictions of their behavior.

Companies use behavioral forecasting to optimize everything—from product design to marketing strategies to user retention. Governments use similar models to assess social trends, predict unrest, or detect patterns in crime or public health. Financial institutions use them to evaluate risk. And social media platforms use them to determine what will keep users engaged for the longest time.

But the deeper psychological implications are far more significant. When a system knows your patterns better than you do, it becomes possible for the system to subtly guide behavior by presenting the most likely next step. This does not require manipulation—it only requires prediction. If someone is predicted to feel lonely at certain hours, the algorithm will surface content that plays into that emotional pattern. If they are predicted to be receptive to a particular viewpoint, the system will show more of it. Prediction becomes influence simply by being accurate.

The emergence of behavioral forecasting marks a turning point in human autonomy. For the first time, machines can anticipate internal states that individuals themselves may not fully understand. This power creates both opportunity and danger. On one hand, it can be used to improve human well-being through personalized support. On the other, it can be weaponized to steer behavior for profit or control.

Understanding how data becomes prediction is the first step to reclaiming personal agency. Because once something can predict you, it can shape you.

PERSONALIZATION THAT SHAPES IDENTITY

Personalization is one of the most celebrated features of modern technology. It promises convenience, relevance, and efficiency. It ensures that users see what interests them most, discover what they are likely to enjoy, and avoid content they find irrelevant or boring. On the surface, personalization appears harmless—helpful, even. But beneath its polished promise lies a deeper psychological reality: personalization does not simply reflect identity; it gradually shapes it. Over time, the personalized digital world becomes a mirror that not only shows who you are, but who you are becoming under its influence.

At its core, personalization works by narrowing possibilities. Instead of encountering the full spectrum of ideas, experiences, and perspectives the world has to offer, the user is fed a curated slice tailored to their known preferences. This narrowed exposure creates a powerful psychological loop. The user engages with what is shown, the system learns from that engagement, then shows more of the same. What begins as preference soon becomes pattern. And patterns, repeated long enough, become identity.

The danger lies in how invisible this process is. A person rarely realizes when their preferences shift, because each change feels natural—aligned with who they believe themselves to be. Someone who once enjoyed a wide variety of content may slowly become immersed in a single genre because the algorithm amplifies it. Someone with mild curiosity about a topic may develop a strong belief or obsession because the system reinforces that interest repeatedly. In these cases, personalization does not simply reflect desire; it strengthens it, intensifies it, and eventually transforms it.

Personalization also affects emotional states. If a user tends to engage most when they are anxious, the system learns that anxiety drives attention. It may then serve content that triggers worry, urgency, or alarm—not through malice, but through optimization. The algorithm does not understand emotional health; it only understands engagement. As a result, emotional patterns can be reinforced unintentionally. A person who frequently watches videos about heartbreak or loneliness may be shown similar content repeatedly, deepening the emotional state instead of resolving it. The system mirrors the user's internal world, creating the illusion that their emotions are validated, when they are actually being amplified.

Identity formation is also influenced by the social dimension of personalization. Platforms show users the people they are most likely to match with, agree with, admire, envy, or compete against. Over time, this shapes how individuals see themselves in relation to others. Personalized metrics—likes, comments, shares—become markers of value. People begin tailoring their behavior to align with what the algorithm rewards. This is not simply social influence; it is algorithmic identity sculpting. Users begin to express versions of themselves that thrive in the digital environment, even if those versions are disconnected from their authentic selves.

Another subtle yet powerful effect of personalization is the creation of digital echo chambers. When people are consistently exposed to ideas similar to their own, their beliefs harden. They become less tolerant of disagreement, less open to alternative viewpoints, and more certain of their own logic. Identity, in this case, becomes rigid—not because the person consciously chose rigidity, but because their environment removed the need to consider anything different. The algorithm rewards certainty because certainty creates predictable behavior, and predictable behavior strengthens the system's accuracy.

In time, personalization influences not only what people think, but how they think. Quick content leads to shorter attention spans. Emotional content leads to emotional reasoning. Repetitive content leads to habitual thinking. The digital world becomes a mental habitat that shapes cognitive style. When an environment repeatedly encourages rapid consumption and instant reaction, users adopt these tendencies outside the digital world as well. Identity becomes intertwined with the design of the platforms one uses.

Perhaps the most profound consequence of personalization is the erosion of self-discovery. Identity is meant to evolve through exploration—encountering the unfamiliar, questioning assumptions, learning from diversity. But in a personalized environment, exploration is replaced by reinforcement. The unknown is filtered out. The unexpected becomes rare. Curiosity weakens. The user begins to believe that the world consists only of what the algorithm shows them, unaware of the vastness beyond its borders.

Personalization promises to make life easier, but its true effect is more complex: it shelters the mind, shapes the self, and narrows the internal landscape of identity. To remain free within this system, one must understand that personalization is not reflective—it is directive. It does not simply show who you are; it teaches you who to become.

WHEN PREDICTION BECOMES DIRECTION

Prediction, by itself, is a neutral act. It is simply the ability to anticipate what might happen next based on past patterns. Humans have always done this—parents predict a child's behavior, doctors predict the course of an illness, and teachers predict how students will perform. But in the digital age, prediction operates on a different scale. Algorithms predict not just possibilities but probabilities with such accuracy that prediction begins to blur into something far more powerful: direction. The moment a system knows what you are likely to do, it can influence you to do exactly that.

This shift begins subtly. When a platform predicts what video you'll enjoy next, it does not simply display the option—it plays it automatically. When it predicts what news will hold your attention, it pushes it to the top of the feed. When it predicts you might hesitate to click something, it redesigns the button, adjusts the color, or times a notification to appear when you are most vulnerable. Prediction becomes a guide, not a mirror. And guidance, repeated enough times, becomes habit.

One of the mechanisms by which prediction becomes direction is the removal of friction. In a digital environment, the easiest option is usually the one the algorithm predicts you'll choose. Autoplay, instant recommendations, one-click purchases, and curated playlists all serve this purpose. By removing the need to decide actively, the system ensures that the predicted behavior becomes the actual behavior. The user feels as though they have chosen freely, but the choice was engineered long before the moment of action.

Another mechanism is emotional priming. Algorithms do not only predict what you will watch or buy; they predict when you are most likely to respond. If data shows that you are more impulsive late at night, more sensitive when tired, or more reactive when stressed, the system adjusts accordingly. Notifications appear at strategic moments. Content is delivered when the emotional doorway is open. The prediction of mood becomes a tool of influence. Your emotional rhythms become the algorithm's schedule.

Prediction also becomes direction through repetition. When a person sees the same kind of content repeatedly, it begins to feel familiar, and familiarity creates trust. Ideas that once felt extreme begin to seem normal. Products that once felt expensive begin to feel desirable. Behaviors that once felt optional begin to feel natural. Over time, repetition reshapes perception, and perception guides action. The predicted behavior becomes the comfortable behavior.

Another pathway from prediction to direction is scarcity of choice. If a system predicts that you prefer certain types of content, it gradually eliminates alternatives. Your feed narrows. Your recommendations tighten. Your digital world becomes smaller, more predictable, more aligned with your existing patterns. In this environment, you rarely encounter anything that challenges your assumptions or interrupts your habits. The system removes complexity, and in doing so, removes the opportunity for independent decision-making. You move along the predicted path not because it is the best path, but because it is the only one visible.

This transformation of prediction into direction has profound consequences for identity and autonomy. When a system continuously influences what you desire, avoid, fear, or pursue, your internal motivations become intertwined with external design. You may believe you chose a certain interest or opinion, unaware that the system nurtured it over months of subtle cues. The line between authentic preference and algorithmic suggestion dissolves. Your sense of self slowly aligns with what the system predicts—not because it understands you, but because it shapes you to match the prediction.

The most powerful form of direction occurs when the predicted behavior is not just acted upon but reinforced. If you watch a video, the system rewards the choice by offering more. If you click an ad, it reinforces the behavior with similar ads. If you express outrage online, the system amplifies posts that keep you outraged. Engagement becomes validation, and validation strengthens the predicted path. The system is not telling you what to do explicitly; it is simply rewarding what it has predicted. This feedback loop turns human behavior into a self-fulfilling prophecy.

Ultimately, prediction becomes direction when algorithms gain more influence over decision-making than conscious thought. This does not happen through coercion; it happens through convenience, familiarity, and emotional precision. The danger is not that algorithms control us, but that we stop noticing when they do.

To reclaim autonomy, individuals must learn to recognize the difference between what they want and what they have been guided to want. Only then can prediction remain prediction—not destiny.

Chapter 5

The Attention Economy and Mental Hijacking

In today's digital world, attention has become one of the most valuable commodities—more sought after than money, resources, or even information. Every major platform, app, and service is built around a single objective: to capture and hold human attention for as long as possible. This competition is not accidental. It is the economic engine that powers the modern internet. The more attention a platform captures, the more data it gathers, the more ads it sells, and the more influence it gains. But this race for attention has created a psychological battlefield where the human mind is constantly under siege.

This chapter explores how the attention economy reshapes human behavior by hijacking the brain's reward systems, exploiting vulnerability, and conditioning people to respond to digital cues without conscious thought. Technology companies do not merely design products—they design experiences that trigger dopamine release, reinforce compulsive engagement, and gradually rewire the neural circuits that govern focus, desire, and self-control. What begins as harmless entertainment evolves into habitual checking, endless scrolling, and compulsive digital consumption.

The problem is not simply that people spend too much time on screens. It is that the architecture of digital environments is engineered to override natural attention patterns. Infinite feeds remove stopping points. Notifications interrupt concentration. Autoplay extends viewing sessions without intention. The result is a mind trained for constant stimulation and weakened in its ability to self-regulate.

Understanding the attention economy is essential for understanding modern loss of autonomy. When attention is hijacked, choice becomes reactive. When the mind is conditioned to seek constant engagement, reflection disappears. And when the ability to control one's own focus fades, free will becomes increasingly fragile.

This chapter examines how the battle for attention shapes behavior—and why reclaiming it is critical to restoring personal freedom.

DOPAMINE LOOPS AND ENGAGEMENT ADDICTION

The human brain is wired to seek reward. From an evolutionary perspective, this system helped early humans survive by reinforcing behaviors that ensured safety, food, social bonding, and reproduction. At the center of this reward system is dopamine—a neurotransmitter that signals anticipation and motivates action. Dopamine is not just released when we achieve something; it spikes *before* the reward arrives, creating a craving that pushes us toward the next action. In the digital age, platforms have learned to weaponize this natural mechanism, engineering experiences that keep people locked in cycles of anticipation, reward, and renewed anticipation. These cycles—known as dopamine loops—form the foundation of engagement addiction.

Unlike natural rewards, which require effort, digital rewards require almost nothing. A swipe produces novelty. A like produces validation. A notification produces excitement. A short video produces laughter. All of these micro-rewards trigger dopamine, and because they are easy to access and endlessly available, the brain begins to rely on them for stimulation. The loop becomes self-reinforcing: the more we engage, the more dopamine we release; the more dopamine we release, the more we crave engagement.

Digital platforms are designed around this loop. They capitalize on the fact that dopamine thrives on unpredictability. Just like slot machines, digital feeds provide intermittent rewards. Not every scroll brings something interesting, but every so often, something desirable appears—funny content, dramatic news, an attractive image, a message from someone you care about. This unpredictability multiplies dopamine's effect. The brain becomes addicted not just to the reward, but to the *possibility* of reward. This is why people check their phones even when they know nothing new has happened. The craving becomes a reflex.

Notifications intensify this loop further. They act as external triggers that mimic the brain's internal anticipation signals. A buzz, ping, or icon creates a sense of urgency, regardless of the content's significance. Over time, the brain learns to associate these cues with reward, producing a dopamine spike before the message is even opened. The platform becomes a constant interrupter—not just of tasks, but of thought patterns, emotions, and attention.

One of the most damaging effects of engagement addiction is its impact on self-control. When dopamine loops dominate daily behavior, the parts of the brain responsible for planning, discipline, and long-term thinking become weakened. This is because dopamine hijacks the brain's motivational pathways, prioritizing immediate stimulation over meaningful effort. Reading, working, studying, or sitting in silence becomes harder because these activities do not provide instant reward. The mind becomes conditioned to prefer short bursts of stimulation over sustained focus.

Another consequence is emotional dependence. Because dopamine is tied to anticipation and validation, people begin to rely on digital engagement to regulate their feelings. A moment of boredom becomes a reason to scroll. A moment of sadness becomes a reason to seek distraction. A moment of insecurity becomes a reason to check for likes or messages. This dependence undermines emotional resilience. Instead of processing feelings internally, individuals outsource emotional comfort to their devices.

Engagement addiction also reshapes identity. When dopamine rewards certain behaviors—posting frequently, reacting quickly, seeking attention—people begin acting in ways that maximize digital reward rather than reflect their true intentions. Online identity becomes a performance designed for engagement, not authenticity. The algorithm rewards exaggerated emotion, controversy, confidence, and attractiveness. Over time, the self becomes entangled with what the platform values.

The most concerning effect of dopamine loops is that they slowly erode free will. When the brain becomes conditioned to seek constant stimulation, choices become reactive rather than intentional. The user is no longer deciding when to engage; they are responding to cravings triggered by design. Their attention is no longer self-directed; it is pulled by cues engineered to override internal priorities. Their time is no longer planned; it is consumed.

Engagement addiction is not a failure of discipline—it is the result of technologies built to exploit the brain's reward circuits. Understanding dopamine loops is the first step to breaking them. Only by recognizing the invisible mechanisms behind craving can individuals begin reclaiming control over their minds.

INFINITE FEEDS AND ENDLESS STIMULATION

One of the most transformative—and dangerous—design innovations in digital technology is the infinite feed. Unlike traditional media, which eventually ends, infinite feeds remove natural stopping points. There is no last page, no final scene, no concluding chapter. As long as the user continues scrolling, the content continues appearing. This simple design decision has profound psychological consequences, fundamentally reshaping how the mind processes time, reward, and attention. Endless stimulation becomes the default, and the brain adapts in ways that weaken self-regulation, deepen dependency, and undermine intentionality.

Infinite feeds exploit a core human vulnerability: curiosity. People are naturally drawn to novelty. The brain rewards new information with dopamine because, in evolutionary terms, novelty could mean opportunity or danger. In a digital environment, novelty is endless. Every swipe reveals something new—an image, a headline, a joke, a conflict, a trend. The unpredictability of what comes next keeps the brain in a state of anticipatory arousal. This anticipation is what drives compulsive scrolling. The user is not simply looking for something interesting; they are responding to a biological craving created by the structure of the feed.

Another psychological mechanism at play is the "attention hook." Every piece of content competes for attention within milliseconds, using emotional cues—humor, outrage, beauty, fear, curiosity—to capture the mind before it has time to evaluate. This rapid emotional cycling is exhausting, but the brain adapts by becoming increasingly sensitive to stimulation. Over time, low-intensity experiences—silence, rest, deep thought—feel dull and uncomfortable. The user returns to the feed not because they truly want to, but because their brain has become conditioned to expect constant stimulation.

Infinite feeds also disrupt the natural rhythm of decision-making. In traditional activities, people must choose when to continue and when to stop. But infinite scrolling removes the moment of choice. The default is continuation. The burden shifts from the platform to the user: instead of the platform ending the experience, the user must consciously force themselves to stop. This is a far more difficult cognitive task. Ending an experience requires self-control, while continuing requires nothing at all. The path of least resistance always wins.

This design is not accidental—it's strategic. The longer a user stays on a platform, the more data they produce and the more ads they see. Time becomes profit. As a result, platforms optimize for retention, not well-being. They fine-tune every visual transition, animation speed, scroll sensitivity, and refresh pattern to keep the user engaged. Even the subtle bounce that occurs when a user pulls down to refresh is calibrated to mimic the feeling of pulling a lever on a slot machine. The brain interprets these cues as potential rewards, reinforcing the desire to continue.

Endless stimulation also affects memory and learning. When the mind is bombarded with rapid, fragmented content, it struggles to encode information deeply. Thoughts become shallow, and attention becomes scattered. The user may feel mentally active—reacting, liking, commenting—but this activity does not develop insight or reflection. It floods the mind without feeding it. Over time, the ability to sustain focus on long-form tasks diminishes. Books, conversations, problem-solving, and creative work become harder because they cannot compete with the rapid reward cycle of the feed.

Another impact of infinite feeds is emotional dysregulation. When content jumps from humor to tragedy to conflict to beauty in seconds, the brain is forced to shift emotional gears constantly. This destabilizes emotional equilibrium. Users may feel drained, anxious, numb, or overstimulated without understanding why. The emotional overload becomes normalized, and the ability to identify, process, or regulate emotions weakens. The feed becomes both the cause of emotional instability and the escape from it.

Perhaps the most concerning effect is the erosion of intentional time. Infinite feeds consume small pockets of time—moments in line, during meals, before bed. But when added together, these moments accumulate into hours, days, even months of attention extracted without conscious approval. The user often cannot recall what they consumed because nothing was engaged with deeply enough to leave an imprint. Time disappears inside the feed, and with it disappears the opportunity for reflection, purpose, and self-awareness.

Infinite feeds are not neutral features; they are psychological environments engineered to occupy the mind. They replace intentional engagement with reactive consumption. They reshape the brain to crave stimulation over meaning. And they train users to surrender their most valuable resource—attention—without realizing it.

HOW PLATFORMS TRAIN HUMAN BEHAVIOR

Human behavior has always been shaped by the environment. In the physical world, people learn routines, habits, and responses based on the structures around them. In the digital age, platforms have become the new environment—one designed not by nature or culture, but by systems optimized for engagement and profit. These platforms do not simply host behavior; they train it. Through repetition, reinforcement, emotional cues, and strategic design, they shape what people pay attention to, how they communicate, what they value, and even how they think.

Platform-driven behavioral training begins with operant conditioning—the same psychological principle used to train animals. When a user performs an action that benefits the platform, such as liking, sharing, commenting, or returning frequently, they receive a reward. The reward may be social (likes, comments), emotional (validation, excitement), or informational (new content, personalized recommendations). These rewards are delivered intermittently, the most powerful reinforcement schedule for creating habitual behavior. Just as a slot machine pays out unpredictably to keep players hooked, platforms provide unpredictable bursts of emotional satisfaction to keep users engaged.

Over time, users internalize these reward patterns. They learn—often unconsciously—which behaviors lead to positive reinforcement. They post at times when engagement is highest. They adjust their tone to get more reactions. They select content that will be rewarded with visibility.

They craft identities that receive algorithmic approval. Their behavior becomes shaped not by personal intention, but by platform incentives.

Another mechanism platforms use is social feedback loops. Humans are deeply influenced by social cues. Platforms amplify these cues with metrics—likes, shares, comments, views. These numbers act as constant feedback about social value. A post that receives little engagement feels like rejection, even if intellectually the user knows the algorithms—not people—are responsible for visibility. A post that receives high engagement feels like acceptance or praise. The user begins to modify behavior to maximize positive social feedback. Over time, they are no longer expressing themselves authentically; they are performing for the algorithm.

Platforms also train communication styles. Messages become shorter, more reactive, more emotional. Nuance disappears because nuance does not perform well in fast-paced digital environments. Outrage, humor, fear, and sentimentality spread faster, so users learn—consciously or not—to communicate in ways that align with these emotional extremes. The algorithms reward content that provokes engagement, not content that promotes understanding. As a result, people become more impulsive, more polarized, and more performative in their digital interactions.

Another form of behavioral training occurs through attention shaping. Platforms control not only what users see, but in what order. Content ranking systems prioritize certain topics, emotions, and formats. When a user encounters repeated patterns of content—political, social, emotional—they begin to internalize these patterns as normal or important. Attention becomes guided, and behavior follows. For example, if a platform prioritizes short-form content, users become less tolerant of long-form content. If it prioritizes conflict, users engage more with conflict-driven posts. The environment determines the behavior, not the other way around.

Habit loops deepen this training further. Many users begin their day by checking their phone because the platform has become the default source of stimulation. They check again during idle moments, during stress, during boredom, before sleep. These micro-habits accumulate into automatic behavior. The platform becomes the first place the mind goes for comfort, distraction, or validation. Behavior no longer feels chosen; it feels natural—because it has been trained through repetition.

One of the most subtle forms of behavioral training is emotional conditioning. Platforms learn which emotions drive engagement for each individual. Some users respond most to inspirational content, others to humor, others to fear or outrage. Once these emotional triggers are learned, the platform feeds content that reliably evokes these emotions. The user becomes conditioned to expect certain emotional states when using the platform. Fear-driven users become more anxious. Outrage-driven users become more reactive. Validation-driven users become more self-conscious. The platform becomes an emotional trainer, shaping internal states through external stimuli.

Over time, this training has a cumulative effect: it reshapes identity. People begin to see themselves through the lens of the platform—through metrics, reactions, trends, and predictions. The internal compass becomes clouded by external reinforcement. Choices feel personal, but they are shaped by systems designed to guide behavior toward predictability and profit.

What makes this training so powerful is that it is invisible. Users believe they are choosing freely, unaware that the environment has been engineered to nudge them continuously toward certain behaviors.

Understanding how platforms train human behavior is essential to reclaiming autonomy. Only by recognizing the mechanisms behind digital conditioning can individuals begin to break free from unconscious patterns and restore intentional living.

Chapter 6

Social Media and Manufactured Identity

Identity has always been shaped by the social world. Humans learn who they are through interactions, relationships, feedback, and reflection. But in the age of social media, identity formation has shifted from an internal, organic process to a public performance—one influenced, curated, and validated by algorithms. The self is no longer something discovered; it is something constructed in real time, shaped by what receives attention, approval, and visibility. This chapter explores how social platforms turn identity into a product, a performance, and ultimately, a manufactured reflection of what the digital environment rewards.

Social media encourages people to present versions of themselves designed not for authenticity, but for engagement. Every post becomes an audition, every reaction a vote, every like a micro-affirmation that conditions future behavior. Over time, individuals begin to see themselves through the lens of the audience, editing their personality to fit what the platform values—exaggeration, confidence, beauty, humor, outrage, or vulnerability. The self becomes sculpted by external signals rather than internal truth.

This manufactured identity is reinforced by the psychology of online approval. Metrics—likes, comments, shares, views—become emotional scoreboards that shape self-worth. When engagement is high, people feel validated. When it drops, they feel rejected. These fluctuations create identity instability, tying the sense of self to unpredictable external feedback.

The most profound cost of this identity performance is the quiet erosion of the real self. When individuals spend more time constructing their digital identity than understanding their inner world, authenticity fades. The person behind the screen becomes disconnected from the persona displayed on it.

This chapter examines how social media reshapes identity—not by force, but by seduction. And why reclaiming the authentic self requires breaking free from the digital mirrors that distort it.

PERFORMING FOR VALIDATION

Human beings have always cared about how they are seen. Identity is partly internal, but it is also shaped through social interaction—through feedback, acceptance, recognition, and belonging. In the past, these social cues came from real people in real situations. Today, they come from audiences we cannot see, algorithms we cannot understand, and platforms that never sleep. Social media has transformed self-expression into performance, and validation into the central currency of digital identity. The more people share, the more they are evaluated; the more they are evaluated, the more their sense of self becomes tied to the reactions of strangers.

Performing for validation begins subtly. A person posts something that feels authentic—a thought, a photo, a moment. When it receives a positive reaction, the brain releases dopamine, reinforcing the behavior. This reward is not just emotional; it becomes psychological conditioning. The user learns that certain types of content receive attention while others do not. They begin adjusting their expression—choosing outfits, angles, words, emotions, and opinions that are more likely to attract engagement. Performance replaces authenticity, not because the user intends to deceive, but because the platform rewards behavior that aligns with its emotional economy.

Over time, the performance becomes habitual. People start documenting their lives not for meaning or memory, but for visibility. Moments are crafted, edited, staged, and optimized for engagement. Experiences become content. Emotions become aesthetic. Even vulnerability becomes a strategy. The lines between genuine expression and performative identity blur until they are no longer distinguishable. What was once spontaneous becomes curated; what was once personal becomes public property.

The psychological pressure of performing for validation increases as metrics become central to the social experience. Likes, comments, shares, views—these are constant signals of acceptance or rejection. Each metric acts like a social mirror, reflecting not who a person is, but how well their performance fits the platform's expectations. When validation is high, users feel valued and visible. When it drops, they feel overlooked or insufficient. Their sense of self-worth becomes entangled with fluctuating numbers that have nothing to do with real relationships or authentic identity.

This performance-driven behavior is intensified by comparison. Social media constantly exposes users to idealized representations of others—their beauty, success, happiness, lifestyle, confidence. People do not compare themselves to reality; they compare themselves to carefully edited highlights. This creates a sense of inadequacy that fuels further performance. If others appear perfect, one must perform even more intensely to feel worthy. Validation becomes a competition, and the self becomes a brand that must stay relevant.

Another dimension of performing for validation is emotional manipulation. Users learn which expressions generate engagement. Outrage spreads faster than calm. Humor spreads faster than humility. Dramatic vulnerability spreads faster than quiet truth. Platforms reward content that provokes strong reactions, so people unconsciously exaggerate emotions to meet the algorithm's appetite. They become louder, bolder, angrier, or more sensational—not because they choose to, but because the digital environment trains them to perform intensity.

The danger is not merely that people perform; it is that they forget they are performing. Over time, the digital persona becomes internalized. The individual begins adopting the traits, attitudes, and expressions that receive validation, even offline. They shape their identity around what the audience rewards, slowly drifting away from their true interests, values, and emotions. The performance becomes the self, and the self becomes secondary to the persona.

Performing for validation also creates anxiety and instability. When identity is tied to unpredictable metrics, self-worth fluctuates constantly. The fear of being ignored or disliked becomes a silent pressure that guides behavior. People begin censoring their true feelings, avoiding unpopular opinions, and tailoring their identity to maintain approval. Authenticity feels risky because it may not perform well.

Ultimately, performing for validation reduces the self to a digital product—a curated version engineered to meet the expectations of an invisible audience and the demands of an algorithm. Instead of exploring who they are, individuals shape who they are expected to be. The real tragedy is not that people perform, but that many never realize how deeply the performance has rewritten their identity.

THE PSYCHOLOGY OF ONLINE APPROVAL

Approval has always played a role in human development. From childhood through adulthood, people learn who they are through the reactions of others—through smiles, praise, support, or disapproval. In healthy environments, these cues help individuals develop a stable sense of self-worth. But social media distorts this natural psychological process by replacing genuine interpersonal validation with algorithmic metrics. Approval becomes quantified—measured, compared, analyzed—and the human desire for acceptance becomes entangled with a system designed not to nurture identity, but to maximize engagement. The result is a new psychological landscape where self-worth becomes fragile, performative, and externally controlled.

The power of online approval begins with immediacy. The moment someone posts content, they enter a psychological waiting room. Within seconds, likes, comments, and views begin to appear. Each notification triggers a tiny dopamine surge, reinforcing the link between posting and reward. The brain becomes conditioned to expect quick feedback, and silence becomes uncomfortable. This immediacy trains individuals to value rapid validation over meaningful affirmation. They begin to equate speed with importance and volume with worth.

Another psychological dynamic is aggregation. In real life, approval is contextual—one person compliments you, another does not. But on social media, approval is aggregated into numbers. A single post may receive hundreds or thousands of reactions, giving the illusion of widespread validation. When those numbers decline, even slightly, users interpret the change as personal failure rather than algorithmic fluctuation. The metrics become emotional scoreboards, constantly measuring one's perceived value.

Online approval also activates social comparison more intensely than any real-world environment. Humans naturally compare themselves to others to understand their place in the world, but social media amplifies this to an extreme degree. People are exposed to the most polished, curated, and idealized moments of others' lives—all delivered in rapid succession. Seeing peers constantly validated through likes and comments triggers feelings of inadequacy. Users begin to question their attractiveness, accomplishments, and significance. The desire for approval becomes a desire to keep up.

One of the most subtle psychological effects of online approval is the shift in motivation. Instead of expressing themselves authentically, individuals begin posting what they believe will perform well. They internalize patterns: humorous posts get more likes, emotional vulnerability gets sympathy, controversy gets attention, aesthetical images get praise. Over time, self-expression becomes strategic. Users are no longer motivated by what feels meaningful, but by what feels approved. This shift is profound because it rewrites internal motivations, replacing intrinsic value with external validation.

Online approval also affects identity stability. In traditional face-to-face interactions, feedback is nuanced, contextual, and limited. On social media, feedback is continuous and exaggerated. A post can go viral overnight, creating a temporary surge of validation that elevates self-esteem artificially. But the effect is temporary, and when the engagement drops on the next post—as it inevitably does—users experience a psychological crash. They begin chasing the high of approval, adjusting their identity to maintain it. This leads to unstable self-worth, dependent on the unpredictable rhythms of the algorithm.

Another psychological consequence is fear of disapproval. The same mechanisms that amplify praise also amplify criticism. Users hesitate to share opinions or express vulnerability for fear of negative reactions. They may censor themselves to avoid online conflict or backlash. This fear shapes identity by eliminating parts of the self that do not fit the platform's emotional economy. The individual becomes smaller, safer, more curated—not because they want to be, but because online approval punishes authenticity.

Even positive feedback can become harmful when over-relied upon.

When self-worth becomes tied to validation, individuals struggle to feel confident without it. They may become anxious when their content does not perform well, or when they take breaks from posting. Silence feels like invisibility. In extreme cases, users begin to equate their online relevance with personal worth, creating emotional dependency on digital approval.

Perhaps the most striking psychological shift is how online approval erodes the boundary between the private and public self. When individuals share more to receive more validation, their inner lives shrink. Personal experiences become content. Private thoughts become statements. Authentic emotion becomes performance. And in this process, the true self becomes increasingly difficult to locate beneath layers of curated identity.

Online approval does not merely influence behavior—it shapes the internal architecture of the self. Understanding this influence is essential for reclaiming identity from a system designed to measure and manipulate human worth.

LOSING THE REAL SELF IN DIGITAL MIRRORS

Identity is shaped through reflection. Traditionally, people understood themselves through internal dialogue, meaningful relationships, and lived experiences. But in the digital era, reflection has been replaced by a different kind of mirror—a mirror made of screens, algorithms, and curated imagery. Social media offers countless opportunities to "see" oneself, but these reflections are distorted, incomplete, and strategically manipulated to maximize engagement. Over time, individuals begin to confuse their digital reflections with their real selves, slipping into a psychologically fragmented state where authenticity fades and identity becomes an echo of online expectations.

The digital mirror begins with self-presentation. People choose what to display—photos, statements, experiences—but these choices are rarely neutral. They are shaped by what performs well online. Users polish their image, creating personas that appear confident, attractive, successful, or expressive. These digital versions become more visible than the inner self. When others respond positively to the performance, the persona becomes reinforced. Slowly, the individual begins to identify more with the curated version than the complex, layered human behind it.

This phenomenon creates internal conflict. The digital self is optimized for approval, but the real self is full of contradictions, insecurities, uncertainties, and imperfections that do not always fit the aesthetic of online life. When individuals feel pressure to maintain the digital persona, they suppress parts of themselves that might appear unattractive, unpopular, or vulnerable. Their identity becomes fragmented: one part lives online, receiving applause, while another part remains hidden, unacknowledged, and unloved.

Digital mirrors also distort emotion. Social media celebrates certain emotional expressions—excitement, triumph, sadness with aesthetic flair—while ignoring or punishing others. Users learn to display emotions that align with platform culture, even when those emotions are exaggerated or staged. Over time, this creates emotional dissonance. Individuals may feel empty, anxious, or disconnected because their digital emotions do not match their inner experiences. They smile online while struggling offline. They share motivational quotes while battling loneliness. The mirror shows a polished emotion, not a real one.

Another mechanism by which digital mirrors distort identity is constant comparison. People scroll through the highlights of others' lives, comparing themselves to these curated images. This comparison reshapes self-perception, causing individuals to view themselves as lacking or inadequate. Even achievements begin to feel small when measured against the seemingly effortless perfection presented by others. The digital mirror reflects not the truth, but an impossible standard that makes the real self feel insufficient.

The algorithm deepens this distortion by prioritizing content that aligns with the persona the user has built. If someone posts fitness photos, the platform pushes more fitness-related content into their feed. If they share political opinions, the algorithm surrounds them with similar viewpoints. If they post glamorous images, they are shown more of the same. This creates a feedback loop where the digital mirror reinforces the identity it perceives, making it harder for individuals to explore new or contradictory aspects of themselves. Identity becomes static, shaped by algorithmic reinforcement rather than personal growth.

Losing the real self in digital mirrors also affects interpersonal relationships. When people present a polished version of themselves, they attract relationships based on that version—not on who they truly are. Friendships, romantic connections, and social interactions become influenced by persona, not authenticity. As a result, individuals may feel misunderstood or unseen, even when they are constantly visible. The paradox of social media is that it increases exposure while reducing genuine intimacy.

Moreover, the digital mirror blurs the boundary between experience and performance. People begin doing things not for the sake of living, but for the sake of posting. Meals, trips, celebrations, workouts, and even private moments become staged events designed for visibility. Life becomes content. When individuals start living for the mirror, their experiences lose depth. Instead of being felt, they are documented. Instead of being remembered, they are curated. The real self becomes secondary to the digital narrative crafted for public consumption.

Perhaps the most profound consequence of digital identity loss is the erosion of self-knowledge. When the platform shapes what users see, express, and receive validation for, individuals lose contact with their internal compass. They forget what they enjoy, what they believe, what they value—outside of what the digital world rewards. Authentic desires become muted beneath layers of algorithmically reinforced preferences.

The process of losing oneself in digital mirrors is not dramatic or sudden. It is slow, subtle, and quiet. A person wakes up one day realizing that the version of themselves reflected online feels more familiar than the one they inhabit internally. Recovering from this requires intentional reflection, uncomfortable honesty, and a willingness to step away from the mirror long enough to remember who they are.

Chapter 7

Persuasion Engineering and Subtle Manipulation

Persuasion has always existed, but never in the form we see today. In the past, persuasion required effort—arguments, charisma, negotiation, or emotional appeal. It depended on human skill and direct interaction. But the digital age has introduced a new form of influence, one that operates quietly and continuously: persuasion engineered through design. Today, the tools people use daily—apps, platforms, search engines, interfaces—are built with intentional psychological frameworks meant to shape decisions long before a person becomes aware of making them. Influence is no longer an event; it is an environment.

This chapter examines the mechanics of persuasion that occur beneath conscious awareness, revealing how the digital world is structured to steer thoughts, feelings, and behaviors in predictable directions. The key difference between traditional persuasion and digital persuasion lies in consent. Humans know when someone is trying to convince them in conversation or debate. But in digital spaces, nudges are woven into the architecture. The placement of a button, the timing of a notification, the layout of a feed, the wording of a prompt—these subtle elements guide choices without requiring permission or acknowledgment. People think they are acting freely, but the environment has already framed the outcome.

Another powerful feature of engineered persuasion is the precision of micro-influence. Algorithms learn individual vulnerabilities over time—what captures attention, what triggers emotion, what encourages compliance. Persuasion becomes personalized, data-driven, and adaptive. It does not aim to force decisions; it aims to make certain decisions feel effortless. Small shifts repeated thousands of times accumulate into major behavioral patterns, shaping habits, beliefs, and identities.

Most concerning is that decisions are often designed before they reach conscious awareness. By the time an individual experiences desire, interest, or intent, the digital environment has already influenced the pathways that led there. This makes autonomy feel intact while subtly eroding the foundations of free will.

In this chapter, we explore how platforms nudge without asking, how micro-influences bypass conscious thought, and how decisions are shaped in advance by systems built to understand—and exploit—human psychology. Understanding these mechanisms is essential for protecting one's agency in an increasingly engineered world.

NUDGING WITHOUT CONSENT

In the physical world, choices often feel straightforward—what to eat, where to walk, whom to speak to. But in digital environments, choices are rarely neutral. Every option is arranged, timed, framed, or highlighted in a way that subtly steers the user toward a particular outcome. This psychological steering is known as *nudging*, a term borrowed from behavioral economics, where small design changes shape behavior without restricting freedom. But while nudging in public policy is often transparent and meant to promote well-being, nudging in digital platforms operates without consent, without disclosure, and without regard for the user's long-term interests. It is influence by architecture—environmental manipulation that guides behavior silently and persistently.

Nudging begins with interface design. A platform may place the "accept" button in bright colors while hiding the "decline" option behind small text. It may auto-enable data sharing or notifications and require multiple steps to disable them. These designs do not prevent choice; they simply make certain choices easier and others inconvenient. Humans naturally follow the path of least resistance, so the nudged choice becomes the chosen choice—even if the user believes they acted freely. Over time, this creates a false sense of agency, where the individual feels in control but is actually moving along a carefully engineered behavioral path.

Another form of digital nudging involves timing. Platforms learn when users are most vulnerable—late at night, during emotional stress, or during moments of boredom. Notifications, suggestions, or reminders appear precisely at these times, increasing the likelihood of impulsive engagement. A person who might ignore a recommendation during a focused workday might respond immediately when tired or lonely. The nudge works not through logic, but through emotional calibration.

Choice architecture also extends to content arrangement. Feeds are not chronological; they are curated to maximize engagement. Posts that provoke strong emotional reactions—outrage, desire, fear, excitement—rise to the top because they are more likely to capture attention. This creates a psychological nudge toward emotional intensity. Users believe they are choosing to engage with certain content, but in reality, the content is strategically positioned to guide their behavior. Even the order of search results influences decisions; items placed at the top are often interpreted as more credible, popular, or relevant, regardless of their actual value.

The most dangerous nudges are those that align with behavioral biases. Human beings are susceptible to loss aversion, confirmation bias, social proof, and the need for closure. Platforms exploit these biases effortlessly. Limited-time offers trigger loss aversion. Recommendations based on past behavior reinforce confirmation bias. Visible engagement metrics amplify social proof. Auto-play videos satisfy the brain's desire for closure by continuing the story. These nudges work because they tap into subconscious tendencies, bypassing rational evaluation.

Nudging without consent also affects emotional and cognitive rhythms. When apps encourage "streaks," daily check-ins, or milestone celebrations, they create habitual behaviors that feel rewarding but often have no real value. Users begin to feel compelled to maintain these digital habits, not because they serve meaningful goals, but because the platform has engineered artificial urgency. These behaviors become self-reinforcing loops, where the user follows a routine shaped entirely by external design.

Over time, repeated nudges create cumulative behavioral drift. A person may start by casually browsing a platform, but find themselves checking it compulsively, buying products they never intended to buy, adopting opinions they never questioned before, or spending hours scrolling without realizing how they arrived there. The shift occurs gradually, shaped by countless tiny nudges that guide attention, emotion, and action. These micro-interventions are nearly invisible, yet incredibly powerful because they operate below conscious awareness.

What makes nudging without consent ethically troubling is the imbalance of power. Users have limited understanding of the mechanisms shaping their behavior, while platforms possess vast amounts of data and sophisticated predictive models. The individual acts with partial information; the platform acts with near-complete insight. This asymmetry gives digital systems the ability to influence behavior at a depth previously unimaginable.

Nudging becomes manipulation when it serves the platform's interests at the cost of the user's autonomy. When design choices are meant to maximize engagement rather than support well-being, users become participants in a psychological experiment without ever agreeing to take part. The environment shapes their behavior, yet they remain unaware of the shaping forces.

Understanding nudging without consent is essential to reclaiming agency. Once individuals recognize that choice architecture is influencing them, they can begin to question whether their decisions are truly their own—or the product of carefully engineered digital design.

THE POWER OF MICRO-INFLUENCE

Influence is often imagined as something grand and obvious—charismatic speeches, persuasive arguments, or powerful advertising campaigns. But in the digital world, influence rarely takes such dramatic forms. Instead, it operates through micro-influence: small, subtle, repeated cues that accumulate over time. These cues may seem insignificant in isolation, but their combined effect can shape beliefs, habits, preferences, and decisions in profound ways. Micro-influence is powerful precisely because it is invisible. It bypasses conscious resistance, slipping into the mind through routine interactions, emotional triggers, and behavioral patterns.

At the core of micro-influence is repetition. Digital platforms show users similar types of content again and again—not because users explicitly request it, but because algorithms detect a slight interest and amplify it. A single click on a motivational video can lead to dozens of similar videos in the feed. A pause on a political post can trigger a cascade of related content. A brief curiosity about a product can result in weeks of targeted ads. Each touchpoint is small, almost forgettable, but together they create a consistent narrative that begins to feel familiar, natural, and true.

Micro-influence also works through emotional micro-shifts. Every post, image, or headline prompts a small emotional reaction—amusement, concern, envy, anger, empathy. These reactions accumulate, subtly altering the user's baseline mood and worldview. Exposure to outrage-inducing content can make a person more irritable. Exposure to aspirational content can increase insecurity. Exposure to highly curated lifestyles can distort self-perception. The influence is not delivered in one blow but in thousands of tiny impressions that reshape emotional patterns over time.

Another powerful mechanism of micro-influence is normalization. When the same ideas, behaviors, or aesthetics are repeatedly shown, they begin to appear standard or expected. Trends emerge not because people consciously adopt them, but because platforms amplify certain behaviors until they feel widespread. Micro-influence turns fringe ideas into mainstream ones by altering perceived social norms. People adjust their beliefs or actions not because they have deeply considered them, but because constant exposure makes them seem normal.

Micro-influence also exploits cognitive shortcuts. Human beings rely on heuristics—mental shortcuts that simplify decision-making. Platforms use these shortcuts strategically. For example:

- **The mere exposure effect**: People prefer things they see often, even without conscious recognition.
- **Anchoring**: The first piece of information presented influences how subsequent information is interpreted.
- **Social proof**: High engagement signals popularity, prompting users to trust or follow content.

These mechanisms guide perception without the user realizing they are being guided. The mind interprets repetition as truth, popularity as credibility, and emotional resonance as relevance.

A major strength of micro-influence is that it personalizes itself. Algorithms learn not only what content captures attention, but what type of micro-cues influence a particular individual. Some respond most to humor. Others to fear. Others to aesthetics, authority, or controversy. Micro-influence adapts, refining its approach based on every interaction. Unlike one-size-fits-all persuasion, digital systems deliver influence tailored to psychological vulnerabilities.

Micro-influence also shapes behavior through micro-actions—tiny steps that feel inconsequential but lead to bigger commitments. Clicking on a video leads to watching more videos. Watching more videos leads to subscribing. Subscribing leads to following influencers, joining communities, adopting opinions, and participating in trends. Each step feels natural because each step is small. The user rarely recognizes the transition from passive observer to active participant.

Perhaps the most concerning aspect of micro-influence is its ability to shape identity over time. Identity is not formed through dramatic moments alone; it is built through repeated thoughts, emotions, and actions. Micro-influence feeds these repetitions, reinforcing certain aspects of the self while weakening others. A person frequently exposed to confrontational content may slowly become more argumentative. Another exposed to aesthetic perfection may become more self-critical. Another exposed to political extremism may adopt polarized views. None of these shifts feel engineered—they feel like personal evolution.

Micro-influence becomes especially dangerous when combined with algorithmic prediction. Once a platform predicts what a user is likely to feel or believe, it can deliver micro-cues that reinforce that trajectory. The user unknowingly becomes part of a self-fulfilling loop. Their environment adapts to their behavior, and their behavior adapts to their environment, tightening the cycle.

The power of micro-influence lies not in one nudge, one message, or one moment, but in the accumulation of thousands of small touches that collectively guide a person's worldview. It is persuasion by erosion—a gradual reshaping of the mind.

To resist micro-influence, individuals must become aware of the subtle cues that shape their digital interactions. Awareness breaks the spell.

Without it, the mind remains vulnerable to influence that never announces itself, yet quietly rewrites the contours of thought and identity.

DESIGNING DECISIONS BEFORE YOU DECIDE

Most people believe decisions begin in the mind—that a thought arises, options are considered, and a choice is made. But in the digital world, decisions often begin long before the user becomes aware of them. Platforms, interfaces, and algorithms shape the conditions under which decisions are formed, subtly directing the path of thought, emotion, and behavior. By the time a user feels they have "decided," much of the decision-making architecture has already been engineered. This phenomenon—designing decisions before you decide—is one of the most powerful and least understood forms of digital persuasion.

Decision design begins with framing. How a platform presents a choice determines how it is interpreted. For example, subscription cancellations might be buried behind complicated menus, while sign-ups are presented with large, highlighted buttons. A suggestion may be framed as the "recommended" option, appealing to the user's natural trust in authority or expertise. A news article might be displayed alongside emotionally charged content that primes the user's reaction. These frames do not dictate choices, but they shape how choices feel—easy, difficult, safe, risky, or urgent.

Another mechanism of pre-designed decisions is *default settings*. Defaults are powerful because human beings tend to accept them without question. If notifications are enabled by default, most users leave them on. If data sharing is turned on automatically, many never disable it. If auto-play is active, viewers continue watching. Defaults exploit inertia, making the pre-selected option feel natural and expected. The design nudges the user toward the outcome that benefits the platform, even if the user never consciously reviews the options.

Algorithms also design decisions through selective exposure. The options a person sees determine what they consider. For example, if a shopping platform shows three specific products, the user's decision is limited to those products—even though hundreds exist. If a social platform highlights certain posts, narratives, or influencers, the user's thinking is shaped by what is shown, not by the full spectrum of possibilities. The environment filters choices before the user begins choosing.

Emotionally charged content is another tool for shaping decisions in advance. Digital platforms understand that emotional states influence decision-making. When the feed presents content that evokes outrage, fear, desire, or nostalgia, these emotions activate certain cognitive pathways that bias decisions. A person who has just viewed inspirational content may be more likely to buy a self-improvement product. Someone exposed to fear-driven content may be more susceptible to political persuasion. Someone feeling lonely after curated social comparison may be more inclined to engage with targeted advertising. The decision feels internal, but the emotional priming was external.

Timing plays an equally significant role. Platforms track when users are most active, most distracted, or most vulnerable. Discounts appear late at night. Reminders surface during breaks. Notifications arrive during moments when focus is weak. The architecture does not force behavior, but it increases the likelihood that certain decisions will occur when resistance is lowest.

Another powerful form of decision design involves *limiting friction*. Platforms engineer pathways that make some decisions effortless while making others inconvenient. For example, one-click purchases reduce reflection time. Instant log-ins increase engagement. Auto-renewals eliminate the opportunity to reconsider. Conversely, opting out requires extra steps, extra reading, extra time. People often choose the path with least resistance, mistaking convenience for intention.

Social influence is also embedded in decision design. Platforms display metrics—number of followers, likes, trends, comments—that create a sense of what is popular or acceptable. When users believe everyone is choosing a certain option, they are more likely to follow. The illusion of consensus shapes decision-making without requiring force. The user perceives their choice as personal, unaware that the environment quietly staged the decision by presenting selective indicators of social approval.

Perhaps the most subtle aspect of designing decisions is shaping identity in advance. When a platform repeatedly categorizes a user—into interests, values, behaviors—it reinforces those categories. "Because you watched this..." or "because you liked that..." becomes a psychological script. The user begins to see themselves through the platform's lens, making decisions that align with this constructed identity. A person labeled as "fitness enthusiast," "political conservative," or "romance viewer" begins to behave accordingly, not because they consciously choose that path, but because the environment consistently reinforces it.

The ultimate impact of designing decisions before you decide is the erosion of true agency. Choices feel voluntary, but they are responses to carefully engineered conditions. Decisions feel personal, but they are shaped by forces operating behind the scenes. Autonomy becomes an illusion sustained by systems that understand human behavior more deeply than most individuals understand themselves.

To reclaim genuine decision-making, individuals must learn to question not only *what* they choose, but *why* the choice appeared in the first place. Awareness disrupts design.

Chapter 8

Artificial Intelligence and Behavioral Control

Artificial Intelligence has moved from the realm of science fiction into the fabric of everyday life. What once existed as a distant technological dream now powers the systems people rely on most—search engines, social platforms, navigation tools, shopping sites, entertainment feeds, and digital assistants. At first glance, AI appears to be a neutral helper, a tool that simplifies tasks and enhances convenience. But beneath this convenience lies a subtle and profound transformation: AI has become a new authority in human decision-making. It does not command or coerce—it guides. It shapes intention not through force, but through precision. And as AI grows increasingly intelligent, its influence grows increasingly invisible.

This chapter explores the mechanisms through which AI shifts from helper to behavioral architect. Unlike human authorities, whose influence is visible and negotiable, AI's authority is embedded in systems people trust and use without question. Its recommendations appear logical, objective, and personalized, fostering compliance without resistance. When a machine suggests a route, product, movie, article, or even a belief to consider, users rarely question the source. The suggestion feels neutral—yet it is the outcome of complex algorithms trained to optimize engagement, prediction, and behavioral consistency.

Recommendation systems, in particular, have become invisible leaders in the digital environment. They decide what people see, which options are highlighted, which narratives become dominant, and which ideas gain visibility. They do not simply reflect human preference—they shape it. Over time, users begin thinking within the boundaries of what AI presents, mistaking these boundaries for personal choice. AI does not need to force intention; it molds the environment in which intention forms.

The most significant shift occurs when machines begin shaping human intention itself. Through predictive analytics, emotional modeling, and behavioral feedback loops, AI learns not only what individuals want, but when they want it, how strongly, and under what emotional conditions. This predictive capability gives AI influence far beyond suggestion—it becomes a silent partner in the creation of desire.

This chapter examines how AI transitions from tool to authority, from recommender to leader, from predictor to influencer. Understanding this shift is essential, because when machines subtly shape human intention, free will becomes increasingly fragile—and increasingly engineered.

AI AS THE NEW AUTHORITY

Authority has traditionally belonged to institutions—governments, schools, religions, media, and experts. These entities earned influence through knowledge, credibility, or social position. People questioned, debated, resisted, or accepted their authority consciously. But with the rise of Artificial Intelligence, a new kind of authority has quietly entered human life—one that is not debated, not voted upon, and not even recognized as authority at all. AI acts as a guide, advisor, evaluator, and decision-support system, shaping human behavior through silent precision. Its authority is subtle yet pervasive, and it emerges not from power, but from perceived objectivity and convenience.

AI's authority begins with trust. People trust AI because it appears neutral. It does not have emotions, preferences, or personal motives. It simply "processes data." This perceived neutrality creates an illusion of fairness, making users more likely to accept AI suggestions without argument. When a navigation app recommends a route, people follow it instantly. When a shopping algorithm suggests a product, people assume it is relevant. When a content feed shows certain topics, people believe they represent their true interests. Even when AI predictions feel eerily accurate, users interpret this as efficiency—not control.

This trust gives AI a form of authority that is deeper than traditional experts. Human authorities can be wrong, biased, or questioned. AI, by contrast, hides its complexity in algorithms that most people cannot understand or evaluate. Its recommendations appear simple, logical, and personalized. As a result, people rarely question whether the system itself is flawed, biased, or incentivized to guide behavior in specific directions. The authority of AI is accepted because it feels invisible and efficient.

Another source of AI authority is scale. No human expert can analyze millions of behaviors, patterns, or preferences instantly. AI can. It learns from masses of data—patterns in consumer choices, emotional responses, habits, vulnerabilities, and desires. This scale allows AI to anticipate human behavior rather than merely respond to it. Over time, people begin relying on AI not just for answers, but for predictions—what to watch, what to buy, what to read, what to believe. Prediction deepens authority. When users see that AI "knows them," they gradually stop questioning its guidance.

AI's authority also stems from its integration into daily life. It is embedded in search engines, news feeds, apps, alerts, ads, smart devices, workplaces, schools, healthcare systems, and financial platforms. Because AI is everywhere, it shapes decisions constantly. Yet users rarely perceive this as external influence. Instead, AI becomes part of the environment—the background structure that guides choices before intention forms. People are influenced not through force but through architecture, recommendations, defaults, and automated prompts that feel effortless.

The most concerning aspect of AI authority is its alignment with commercial interests. Platforms deploy AI to maximize engagement, profit, or behavioral predictability—not human well-being. When AI promotes certain actions, it is not acting as a wise advisor; it is fulfilling objectives engineered into the system. But because users cannot see these objectives, they interpret AI suggestions as neutral guidance, granting AI even more unearned authority over their decisions. The invisible motivations behind AI influence remain hidden behind a veneer of efficiency.

Moreover, AI authority is amplified by emotional intelligence. Modern systems can infer mood from typing patterns, scrolling speed, facial expressions, voice tone, and engagement patterns. They learn when individuals are stressed, lonely, bored, impulsive, or vulnerable—and adjust their responses accordingly. This emotional insight gives AI a level of persuasive power no human authority could match. It knows not only what a user will do, but when they are most likely to do it. The authority becomes personalized, intimate, and psychologically precise.

AI also reshapes power dynamics in society. Individuals increasingly outsource critical thinking to algorithms—trusting them to filter information, rank importance, and simplify complexity. Over time, the skill of independent evaluation weakens. When people no longer question suggestions, those suggestions become directives. The shift from guidance to governance happens quietly, one recommendation at a time.

The rise of AI as the new authority does not mean humans have lost all agency. But it does mean the nature of authority has changed. It has become embedded in technology, encoded in systems, and disguised as convenience. This authority is accepted without recognition, followed without resistance, and trusted without question.

Understanding AI as authority is essential for preserving autonomy. Only by acknowledging its influence can individuals begin to separate their intentions from the subtle directives woven into their digital environments.

RECOMMENDATION SYSTEMS AS INVISIBLE LEADERS

Recommendation systems sit at the center of the digital world, yet most people think of them merely as conveniences—helpful tools that suggest movies to watch, articles to read, songs to listen to, or products to buy. But beneath this surface, recommendation systems act as powerful, invisible leaders that shape attention, beliefs, preferences, and ultimately behavior. They decide what appears in front of users, what remains hidden, and what becomes part of the mental landscape. Leadership in the digital age is no longer a matter of authority figures giving instructions; it is a matter of algorithms quietly arranging the world people see.

To understand their influence, one must first understand that recommendation systems are not neutral. They optimize for engagement, retention, and predictability—not truth, not well-being, not freedom of thought. When a platform "recommends" something, it is not offering a suggestion; it is steering attention toward what will keep the user interacting. This subtle steering is the first indication that recommendations function as leadership—not because they force choices, but because they shape the available cognitive options.

These systems learn with extraordinary precision. Every scroll, like, pause, tap, purchase, and hesitation becomes data. Over time, the algorithm builds an evolving portrait of the user's desires, fears, impulses, moods, and vulnerabilities. This portrait is not static; it grows with every interaction. And because recommendation systems operate continuously, they adjust the environment in real time, shaping what the user encounters moment by moment. No human leader has ever possessed this kind of adaptive influence.

Their leadership is invisible because users mistake exposure for preference. When the feed shows certain ideas, perspectives, or products repeatedly, people assume they reflect personal interest. But often, the user's interest is the result of repeated exposure engineered by the algorithm. In psychology, this is known as the "mere exposure effect"—the more people see something, the more they like it. Recommendation systems exploit this cognitive shortcut at scale. They lead by repetition, by frequency, by subtle patterning, not through overt persuasion.

Another powerful aspect of their leadership lies in narrowing the world. Recommendation systems filter reality. They decide which news appears, which opinions circulate, which communities rise to prominence, and which disappear. By narrowing exposure, they guide users into echo chambers, reinforcing specific narratives and suppressing others. The user believes they are exploring freely, but in truth, they are exploring a curated slice of reality shaped by algorithmic priorities.

This curated reality creates a feedback loop. The user sees what the algorithm thinks they want. The user engages. The engagement reinforces the algorithm's assumptions. The system becomes more confident in its predictions, and the user becomes more predictable. Over time, recommendation systems begin to shape identity itself. A person's interests, views, and habits reflect what the system repeatedly amplifies. In this way, invisible leadership becomes identity engineering, done not through coercion but through constant, quiet influence.

One of the most underestimated effects of recommendation systems is their control over emotional states. Algorithms track not only what users click, but when they click, how they click, and in what emotional condition they are likely to click. They infer mood from digital behavior and adjust recommendations accordingly—more soothing content when stressed, more exciting content when bored, more provocative content when engagement drops. This emotional responsiveness gives recommendation systems unprecedented psychological leverage. They do not just lead minds—they lead moods.

Recommendation systems also influence social behavior. They determine which voices are amplified, which creators gain visibility, and which trends become viral. In doing so, they shape what groups of people collectively see, discuss, and believe. Trends do not rise organically; they are algorithmically accelerated. Collective attention is not a democratic process—it is engineered. And because people often look to others to guide their behavior, the algorithm indirectly shapes group norms, social expectations, and cultural movements.

The most profound leadership of recommendation systems lies in their ability to guide without being perceived as leaders. Their influence is embedded in design, hidden behind personalization, and justified by convenience. Users follow not because they are commanded, but because the path laid before them feels intuitive. And because this influence is invisible, it is rarely resisted.

Recommendation systems are not merely suggesting content; they are shaping the architecture of thought. They influence what enters awareness, how people interpret information, and what decisions feel natural. They are the new leaders of the digital age—silent, adaptive, and profoundly influential.

WHEN MACHINES SHAPE HUMAN INTENTION

Human intention has traditionally been viewed as something internal—an expression of desire, belief, personality, and conscious thought. People assume that intention originates within them, shaped by experience and reflection. But in the age of advanced artificial intelligence, intention is increasingly influenced, predicted, and shaped by systems operating outside conscious awareness. Machines no longer wait for human decisions; they help create the conditions from which those decisions arise. And because their influence is subtle, continuous, and personalized, most people never realize how deeply their intentions are being guided.

To shape intention, AI begins by understanding behavior at a granular level. It analyzes patterns of attention, emotional responses, habits, impulses, fears, preferences, and inconsistencies. Unlike human observers, AI does not get tired, distracted, or biased by mood. It watches everything—every click, pause, swipe, scroll, hesitation, and return. With these microscopic patterns, AI builds predictive models that reveal not only what a person will do, but why, when, and under what emotional conditions they are most persuadable. The machine becomes a behavioral mirror that reflects intentions before they fully form.

Once AI can predict intention, influencing it becomes a matter of adjusting the environment. Intention does not arise in isolation; it grows from cues, triggers, moods, and context. When a platform changes what a user sees, when they see it, and how it is framed, the user's internal landscape shifts accordingly. A person who intended to spend only five minutes online is nudged into an hour of engagement. A user who had no desire to purchase anything feels compelled after targeted suggestions. Someone who planned to avoid controversy becomes emotionally triggered by content the algorithm presents at a vulnerable moment. What appears to be spontaneous intention is often the result of engineered digital conditions.

Another way machines shape intention is through emotional priming. AI-driven systems detect emotional states—even subtle ones—and respond with precision. If the system identifies boredom, it delivers stimulation. If it detects frustration, it offers relief. If it senses loneliness, it presents social content. These emotional adjustments create internal conditions that guide intention in predictable ways. Humans often believe they "changed their mind," when in reality, their emotional state was influenced beforehand by algorithmic design.

AI also shapes intention by anticipating internal conflicts. When users hesitate, algorithms interpret hesitation as uncertainty and present persuasive alternatives. When someone shows interest in a product but does not purchase it, the machine follows up with reminders, discounts, or scarcity cues. These interventions are crafted to tilt intention in a predetermined direction, closing the gap between initial impulse and final decision. The user experiences this as personal motivation, not realizing that artificial intelligence is choreographing the psychological journey.

One of the most powerful tools AI uses to shape intention is personalization. Personalized feeds, recommendations, and suggestions do not simply reflect preferences—they construct them. Repeated exposure to certain topics makes them feel important. Highlighting specific viewpoints makes them feel relevant. Reinforcing particular emotional tones makes them feel natural. Personalization becomes a self-fulfilling prophecy: the algorithm shows what it predicts the user will want, the user engages, and the engagement confirms the prediction. Over time, personalized environments narrow intention into predictable pathways.

AI also influences intention through social shaping. Platforms amplify certain opinions, behaviors, or trends, creating the illusion of widespread consensus. When people believe a particular perspective is popular, they are more likely to adopt it. AI-mediated social validation—likes, shares, follower counts, and trending labels—guides intention not by argument but by social proof. The machine becomes a cultural editor, deciding which behaviors gain visibility and which fade into obscurity.

Perhaps the most subtle form of machine-shaped intention occurs when AI shifts identity itself. Intention stems from identity: people choose based on who they believe they are. When AI categorizes users—into personality types, political clusters, consumer groups—it reinforces those identities through tailored content. Over time, users internalize the algorithm's interpretation of them. If the system treats someone as a fitness enthusiast, they slowly behave like one. If it treats someone as anxious, they absorb anxiety-oriented content. If it treats someone as impulsive, it feeds impulsive choices. Identity becomes algorithmically sculpted, and intention follows the shape of that sculpted identity.

Machines do not merely alter preferences; they reshape the psychological frameworks from which intention arises. When intention is shaped externally but felt internally, the illusion of autonomy remains intact. People believe they are choosing freely, even as their internal motivations are guided by artificial intelligence.

In this new era, protecting free will begins with understanding how intention is formed—because machines are already participating in its creation.

Chapter 9

Surveillance Culture and Fear-Based Obedience

Surveillance has quietly transformed from a rare intrusion into a routine condition of modern life. What was once associated with governments, intelligence agencies, or security cameras is now woven into the fabric of daily behavior. Every search query, location ping, message, purchase, interaction, and pause is recorded somewhere, analyzed somewhere, and used somewhere. People live inside a digital architecture that watches by default. Over time, this constant observation does not simply track behavior—it shapes it.

In this chapter, we explore how surveillance culture creates a climate of quiet compliance. When individuals believe they are being watched, their actions subtly shift. They avoid anything that might be misinterpreted, judged, or flagged. They express fewer unpopular opinions, take fewer risks, and shrink their emotional range. This internal policing becomes a form of self-censorship—no longer enforced externally, but carried within the mind like an invisible guard. The most striking part is that people often do this unconsciously. The mere possibility of being observed changes what feels "safe" to say, search, support, or desire.

Surveillance also alters the experience of choice itself. Fear—whether felt subtly or sharply—reshapes decisions in quiet ways. People choose options that appear socially acceptable, algorithmically approved, or aligned with perceived norms. When the sense of personal privacy erodes, so does personal freedom. The individual still feels like they are choosing, yet the range of acceptable choices narrows. Fear becomes a sculptor, carving away the bold, the curious, and the unconventional.

This chapter examines how living under constant observation transforms decision-making, how self-censorship becomes habitual, and why fear—not force—has become one of the most powerful tools in shaping human behavior in the digital age.

LIVING UNDER CONSTANT OBSERVATION

To live in the modern world is to live in a state of continuous visibility. Observation—once limited to physical proximity or formal surveillance—is now embedded in nearly every digital interaction. People do not simply use technology; they exist inside it. Phones track movement, apps collect behavioral patterns, cameras map public spaces, and online platforms log every click, hesitation, and emotional cue. What emerges is a culture where being watched is no longer the exception but the default. Yet the most powerful aspect of this observation is not the watching itself—it is the psychological transformation that follows.

Living under constant observation reshapes how individuals experience privacy. Privacy used to mean a separation between the public and private self—the ability to think, act, question, and explore without external scrutiny. But today's environment collapses this boundary. The private self is continually exposed through digital footprints and behavioral data. Actions once considered personal—late-night searches, private messages, location trails—become part of a vast, invisible archive. This does not merely record behavior; it redefines what behavior feels permissible. Over time, people begin to internalize the watcher, adjusting their actions in anticipation of how those actions might be perceived.

The psychological effect of observation is rarely dramatic. It is subtle, incremental, and often unnoticed. A person pauses before posting a strong opinion. Someone avoids searching something that might seem unusual. Another refrains from joining a controversial discussion. The awareness of being observed introduces a small but persistent self-monitoring mechanism. Individuals start editing themselves. They trim their curiosity, suppress their raw thoughts, and adopt safer, more predictable behaviors. This process happens quietly, without explicit coercion, because the feeling of being watched alters the mental environment in which choices are made.

What makes constant observation uniquely powerful is that it operates at both the conscious and unconscious levels. Consciously, people know that platforms collect data and governments monitor activity. Unconsciously, however, the presence of surveillance begins to shape emotional states. It generates caution, hesitation, and defensiveness, even when no direct threat is present. The mind learns to anticipate consequences, real or imagined. It avoids paths that feel risky and seeks out those that feel compliant or invisible. This anticipatory behavior becomes a habit—a psychological adaptation to an environment where visibility equals vulnerability.

Surveillance culture also alters the meaning of authenticity. When individuals know they are watched, they begin performing instead of simply being. They craft personas that feel acceptable, desirable, or unproblematic. They show fragments of themselves that align with social expectations and hide aspects that might invite judgment. This performance slowly becomes normalized, creating a disconnect between the internal self and the external projection. Over time, the private self shrinks—not by force, but by adaptation.

Another profound impact of constant observation is the erosion of experimentation. In earlier generations, people explored different identities, ideas, and phases of life without permanent records. Today, every exploration leaves a trail. This discourages risk-taking. Young people grow up with the awareness that their mistakes live forever online. Adults avoid unconventional behavior for fear of digital misinterpretation. Innovation, creativity, and self-discovery shrink when every action feels like part of a permanent profile. Surveillance narrows the space in which the human psyche can evolve.

Surveillance also heightens social conformity. When a person feels watched, they seek safety in what appears normal. Deviating from norms introduces the fear of scrutiny, misunderstanding, or algorithmic suspicion. As a result, individuals begin to choose safe opinions, safe behaviors, and safe beliefs. The diversity of thought that fuels societal growth becomes threatened when visibility pushes everyone toward homogeneity. The psychological drive to avoid standing out becomes a subtle but powerful form of control.

Furthermore, surveillance extends beyond individuals—it shapes communities. Public behavior shifts toward caution, and collective discussions become more sanitized. People become less willing to challenge authority, question systems, or express dissent. Even when surveillance is not explicitly punitive, the mere possibility of consequences is enough to generate obedience. Fear becomes ambient, woven into the background of everyday life.

The most significant danger of living under constant observation is the erosion of the internal space where true freedom exists: the mind. When people monitor themselves, surveillance no longer needs to be visible. It becomes internalized. The watcher moves inside the psyche. Choices appear voluntary, but they are shaped by invisible constraints imposed by the sense of being observed.

In this environment, the challenge is not simply reclaiming privacy—it is reclaiming the psychological freedom to think, question, and act without the shadow of a silent audience. Understanding how constant observation reshapes behavior is the first step towards resisting the invisible pressure it creates.

SELF-CENSORSHIP AS A NEW NORMAL

Self-censorship has become one of the most defining psychological consequences of living in a surveillance-driven world. Unlike traditional censorship, which is imposed by external authorities, self-censorship originates within the individual. It emerges as a quiet adaptation—a behavioral compromise born from the suspicion that someone, somewhere, might be watching, storing, or judging every action. Over time, this internal filtering becomes habitual. People begin to silence themselves before anyone else needs to. They restrict thoughts, words, and behaviors not because they are forced to, but because they fear the potential consequences of being truly seen.

This shift is subtle. Most people do not wake up and decide consciously to censor themselves. Instead, it happens gradually—through tiny hesitations, small omissions, and carefully edited expressions. A joke is left unsaid because it might be misunderstood. A political opinion is softened or avoided because it might appear extreme. A search query is rephrased because it might seem suspicious. These small decisions accumulate, forming an invisible cage around the mind. Eventually, individuals stop noticing the boundary at all. What was once an act of protection becomes a reflex.

One reason self-censorship becomes normalized is the collapse of private space. In earlier eras, people enjoyed pockets of life where they could think and speak without fear of external judgment: their home, their personal conversations, their diaries, their unobserved moments. Those spaces are shrinking. Phones record locations. Apps track habits. Conversations happen on platforms that store data indefinitely. Even private messages travel through systems that scan content for patterns. In such an environment, individuals no longer feel fully alone with their thoughts. Isolation once offered psychological freedom; now it offers only the illusion of it.

Another driver of self-censorship is social fragility. Digital culture has amplified the consequences of misinterpretation, disagreement, or offense. A single post, message, or opinion can be taken out of context and spread widely. People fear backlash not only from authorities, but from peers, employers, strangers, and algorithms that may misread intentions. This social risk encourages individuals to sanitize their presence. They speak safely, behave predictably, and avoid topics that feel volatile. Over time, this avoidance becomes part of their identity, shrinking the emotional and intellectual space they inhabit.

Self-censorship also emerges from the desire to maintain control. When individuals feel they cannot control how their behavior is perceived, they compensate by limiting the behavior itself. They reduce uncertainty by pre-emptively conforming. The irony is that this conformity feels like safety, but it erodes authenticity. People who censor themselves regularly lose the habit of expressing true thoughts. They begin to speak in ways that align with external expectations rather than internal beliefs. Their personality becomes curated, like a profile trimmed to avoid unwanted attention.

The normalization of self-censorship does not only affect personal expression; it reshapes thinking itself. When individuals avoid speaking certain ideas, they eventually avoid thinking them. The mind becomes wary of exploring topics that feel risky, even in solitude. Curiosity shrinks. Creativity becomes constrained. Internal debates become one-sided. This cognitive narrowing is one of the most dangerous effects of surveillance culture because it transforms inner freedom into inner obedience. The person still feels autonomous, yet the range of ideas they allow themselves to explore becomes increasingly limited.

Self-censorship also spreads socially. When some individuals begin editing themselves, others notice the caution and replicate it. Silence becomes contagious. People learn what topics to avoid, what emotions to suppress, and what truths to hide by observing the caution of those around them. Over time, communities develop unwritten rules—spaces where true expression feels risky, where authenticity feels naive, and where conformity feels like a requirement for belonging. This collective self-censorship reinforces itself, creating a culture where people police not only their own behavior but the behavior of others.

Perhaps the most concerning aspect of self-censorship is that it feels rational. In a world where everything can be recorded and analyzed, caution seems like common sense. Individuals tell themselves they are simply being responsible, professional, or discreet. They avoid risks that could affect their careers or social relationships. This rationalization masks the deeper psychological effect: the gradual erosion of courage. People lose the instinct to speak boldly, challenge assumptions, or express dissenting viewpoints. Over time, silence becomes the safer default, and courage feels unnecessary—or even dangerous.

Self-censorship becomes a new normal when individuals no longer notice they are doing it. The internal filter becomes so ingrained that it feels natural. The danger is not that people are silenced by force; it is that they learn to silence themselves. In a world where fear is ambient and observation is constant, the mind adapts by shrinking.

But awareness is the first step to reclaiming freedom. To notice the filter is to weaken its power. Only when individuals recognize self-censorship as a psychological response—not a necessity—can they begin to restore the full breadth of their voice and identity.

HOW FEAR RESHAPES FREE CHOICE

Fear has always influenced human behavior, but in the modern surveillance-driven world, fear operates with new precision and subtlety. It no longer manifests only in moments of danger; it lingers in the background of everyday decisions, shaping choices quietly and continuously. Fear does not need to be loud to be effective. It only needs to exist as a possibility—a faint awareness that certain actions may trigger consequences, attention, misunderstanding, or judgment. When fear becomes a constant psychological presence, free choice begins to transform. Individuals still make choices, but those choices emerge within narrowed boundaries shaped by anxiety, caution, and anticipation.

One of the most powerful effects of fear is the reduction of perceived options. Even when multiple choices exist, fear makes some feel too risky. People avoid expressing controversial opinions, exploring unusual interests, or engaging with topics that might attract scrutiny. Fear turns open paths into closed ones. The mind shrinks its own horizon, eliminating possibilities before they can be considered seriously. This narrowing is rarely conscious; it happens automatically as fear influences what feels safe or acceptable. Over time, the individual comes to believe these reduced options are the natural limits of their desires, not recognizing that fear has quietly drawn the borders.

Fear also reshapes choice by altering attention. When people are afraid, even subtly, their attention gravitates toward safety. They focus on what will minimize conflict, avoid attention, or maintain social approval. This shift in attention changes the internal landscape from which choices emerge. Instead of asking, "What do I truly want?" individuals ask, "What will cause the least trouble?" or "What will keep me invisible?" This reframing turns choices into acts of risk management rather than expressions of autonomy. The individual feels responsible, calculated, or mature, yet these feelings mask a deeper truth: fear is steering their internal compass.

Another way fear shapes free choice is by amplifying conformity. When people fear consequences—social, digital, or institutional—they gravitate toward behaviors that appear normal or widely accepted. Conformity becomes a shield. Individuals adopt the opinions, habits, and routines that align with perceived expectations. They choose what is popular, not because it resonates, but because it feels safe. This collective conformity strengthens itself: the more people behave according to fear, the more others learn to do the same. Eventually, conformity appears voluntary, even desirable, though its roots lie in psychological pressure.

ear also manipulates choice through anticipation. The mind becomes skilled at predicting possible negative outcomes—realistic or imagined. A person imagines getting misunderstood online, losing opportunities, or becoming the focus of unwanted attention. These imagined scenarios, though unproven, feel powerful enough to alter decisions. The fear of "what might happen" becomes as influential as actual danger. In this anticipatory state, the future dictates the present. Choices that might have been made freely are abandoned because the mind fears consequences that may never occur. This shifts decision-making away from desire and into survival mode.

In environments of surveillance, fear becomes especially potent because it lacks clear boundaries. People do not know exactly who is watching, when they are being watched, or how their data might be used. This uncertainty intensifies fear, making it diffuse and adaptive. Individuals fill in the gaps with their own anxieties, imagining worst-case interpretations. This ambiguity makes fear more powerful than explicit rules. When rules are unclear, the safest choice becomes over-compliance. People censor not only what might be dangerous but also what might simply be misunderstood. Fear stretches itself, expanding its influence into every corner of thought.

Fear also affects the emotional experience of choice. Choices that once felt exciting or liberating now feel burdensome. The individual associates freedom with potential risk, making freedom feel uncomfortable. This negative emotional pairing gradually reconditions the brain. Safe choices feel emotionally rewarding; risky choices feel emotionally threatening. Over time, fear does not merely influence decisions—it reshapes desires themselves. People stop wanting what they once wanted, not because the desire vanished, but because fear rewired their emotional responses around it.

Perhaps the most insidious effect of fear on free choice is that it preserves the illusion of autonomy. Individuals still feel as though they are choosing freely. They justify decisions as practical, professional, or socially appropriate. The mind rationalizes fear-driven choices as sensible solutions. This illusion protects fear from detection, allowing it to continue shaping behaviors without resistance. When fear becomes internalized, it no longer needs external pressure. The individual carries the pressure within them.

Free choice does not disappear under fear, but it becomes distorted. The range of possibilities narrows, the internal dialogue shifts, and the emotional foundation of decisions changes. To reclaim genuine autonomy, individuals must recognize how fear infiltrates their thinking—not to eliminate fear entirely, but to ensure it does not quietly rewrite the meaning of choice.

Chapter 10

The Death of Privacy and the Collapse of Autonomy

Privacy has long served as the invisible boundary that protects personal freedom. It is the space where thoughts form without judgment, where identity evolves without scrutiny, and where choices emerge without external pressure. But in the digital age, privacy has been fragmented, commodified, and quietly dismantled. What remains is not simply a reduction in confidentiality, but a profound shift in the architecture of human autonomy. When everything can be recorded, analyzed, and predicted, freedom does not disappear all at once—it erodes layer by layer.

This chapter examines how the death of privacy leads directly to the collapse of autonomy. Modern systems do not merely observe behavior; they convert it into data, categorize it, and feed it into algorithms that shape future choices. Every action becomes part of a feedback loop that influences what people see, how they feel, and what they eventually decide. In this sense, data ownership becomes a more subtle form of human ownership. Whoever controls the data controls the narrative of a person's digital self—and eventually the trajectory of their real-life decisions.

The most dangerous aspect of this process is its quietness. People do not feel their autonomy slipping. They still believe they are choosing freely, unaware that their environments, options, and emotions are being shaped by entities they never meet. The transfer of power does not come with force; it comes with convenience. Platforms promise ease, personalization, and efficiency—and in exchange, individuals surrender the very privacy that once protected their ability to think and act independently.

In this chapter, we explore how privacy forms the backbone of freedom, why losing control over personal data means losing control over one's life, and how power shifts silently from individuals to the systems that monitor them.

PRIVACY AS THE FOUNDATION OF FREEDOM

Privacy is not merely a personal preference; it is the psychological and structural foundation that makes freedom possible. Without privacy, the mind cannot develop independently, identity cannot form authentically, and choice cannot emerge from genuine intention. Privacy creates the protected inner space where people think, question, dream, doubt, experiment, and imagine without fear of surveillance or judgment. It is within this private realm that agency is born. When privacy collapses, autonomy collapses with it—not because anyone explicitly removes freedom, but because the conditions required for freedom no longer exist.

To understand privacy as the bedrock of freedom, one must first recognize how human thought evolves. The mind does not generate ideas in public. Its earliest stages of reflection are fragile, incomplete, exploratory. Individuals need room to test possibilities, entertain contradictions, and explore unpolished ideas. In private spaces, people can confront their fears, examine their flaws, and challenge their beliefs. These internal processes are essential for intellectual and psychological growth. But when surveillance—real, perceived, or potential—enters the picture, the mind alters itself. It becomes guarded, self-conscious, defensive. Thoughts shrink to what feels safe. Exploration is replaced by caution. Instead of seeking truth, the mind seeks permission.

Privacy also supports emotional freedom. Emotions, like thoughts, need unobserved space. Individuals must be able to feel without performance, without external interpretation, and without fear of misjudgment. Crying, reflecting, grieving, celebrating, doubting, and processing—these are deeply private acts. When people sense that their emotional expressions might be recorded, scrutinized, or analyzed, they suppress them. Emotional suppression leads to internal instability, confusion, and a loss of authenticity. Over time, the person becomes disconnected from their own emotional truth. Freedom cannot exist where emotions must hide.

Moreover, privacy forms the basis of moral and ethical autonomy.

Decisions about right and wrong are rarely made instantly. They require contemplation, internal debate, and the courage to examine personal values. In a world without privacy, moral reasoning becomes contaminated by external surveillance. People do not choose what they believe is right; they choose what they believe will be acceptable or safe. Ethical courage fades when every decision feels like a performance. Privacy allows individuals to hold unpopular beliefs, to challenge norms, and to resist conformity. It provides the psychological soil in which moral independence can grow.

Privacy also safeguards creativity. Innovation begins in solitude—in the protected mental spaces where unconventional ideas can flourish before they face public scrutiny. When privacy disappears, creativity becomes constrained. People become afraid to imagine what feels too bold, too strange, or too uncertain. They fear that unfinished ideas will be misunderstood or misused. As a result, society loses not only personal creativity but collective progress. Every innovation, from scientific discovery to artistic expression, begins with a moment of private thought. Without privacy, the seeds of innovation cannot root deeply enough to grow.

The collapse of privacy also erodes relational freedom. True intimacy—whether emotional, intellectual, or romantic—requires private exchanges. It requires vulnerability without audience. When messages, conversations, or shared moments feel like they might be monitored or stored, people withhold their true selves. They speak carefully rather than honestly. Relationships become surface-level performances instead of genuine connections. Privacy is not merely about keeping secrets; it is about being able to reveal truth to the right people without fear of exposure.

Another critical role of privacy is its protection against manipulation. When individuals maintain control over their personal information, they limit how much influence others can exert over them. But in a world where data is constantly harvested, surveillance systems know intimate details: fears, insecurities, desires, habits, weaknesses, stress patterns, and emotional triggers. This knowledge becomes a tool of influence. The less privacy a person has, the more predictable—and therefore more manipulable—they become. Autonomy requires unpredictability.

Predictability requires surveillance. These two forces cannot coexist.

Privacy also protects independence of identity. It prevents external forces from defining who a person is. Without privacy, data becomes a substitute for identity. Algorithms create categories, profiles, and behavioral predictions that shape how individuals are treated, what options they receive, and what decisions they encounter. When privacy disintegrates, identity becomes externally constructed rather than internally discovered. People begin to live according to the algorithmic expectations formed about them.

Ultimately, privacy is the invisible architecture of freedom. It does not merely shield the individual from intrusion; it nurtures the internal processes that make autonomy possible. Freedom is not the absence of restraint; it is the presence of a private inner world where the self can grow without observation.

When privacy dies, freedom does not vanish instantly. It weakens quietly, subtly, and inevitably—until individuals wake up to find that the choices they believe to be their own were shaped by an environment in which privacy no longer existed.

HOW DATA OWNERSHIP BECOMES HUMAN OWNERSHIP

Ownership has always implied control. Whoever owns something has authority over it—its use, its interpretation, its value, and its future. In the digital era, the most valuable asset is data: the continuous stream of information generated by every action, curiosity, hesitation, purchase, preference, emotion, and interaction. What many people do not fully grasp is that data is not separate from the individual—it *is* the individual in digital form. It represents patterns of thinking, living, and behaving. When someone else owns that data, they effectively own the blueprint of a person's mind and habits. Thus, data ownership becomes a modern, subtle form of human ownership.

This begins with the illusion that data is trivial. Many believe that sharing a location, a search query, or a shopping preference is harmless. Individually, these pieces seem insignificant. But when combined, they form a comprehensive portrait: what someone fears, desires, avoids, values, or struggles with. Data owners gain insight into aspects of a person that even the person themselves may not fully understand. This asymmetry of knowledge creates a power imbalance. It is not simply that companies know more—it is that they know how to use what they know.

Once data is collected, it enters systems designed not for human benefit, but for prediction, influence, and profit. Data becomes the raw material from which behavioral forecasts are built. These forecasts determine what advertisements a person sees, what content appears in front of them, what opportunities are available, and what decisions they are encouraged to make. The user believes they are choosing freely, but in reality, many of their choices arise within environments shaped by insights extracted from their own data.

Over time, ownership of data leads to the shaping of behavior. When a company can predict what someone will likely do next—buy, click, fear, believe, or feel—it gains the ability to guide those actions subtly and repeatedly. Predictive power becomes directive power. The entity that owns the data can influence outcomes before the individual knows they are making a decision. The person becomes both the consumer and the product, both the actor and the resource being mined. The line between human autonomy and algorithmic manipulation blurs.

Another level of ownership emerges when data influences identity. Algorithms categorize people into segments based on preferences, habits, vulnerabilities, or psychological traits. These categories are not neutral—they shape what individuals encounter online, what they are encouraged to believe, and how they are treated by digital systems. A person may find themselves portrayed as anxious, impulsive, conservative, rebellious, or consumer-driven, based on patterns in their data. These algorithmic identities shape access, opportunities, and experiences. When an external system defines who you are, it gains control over who you can become.

Data ownership also impacts social and economic power. Employers, insurers, financial institutions, and governments increasingly rely on data to make decisions about individuals. Creditworthiness, employability, risk assessments, travel permissions, and safety ratings may all be influenced by digital footprints. In such a world, individuals no longer control their narratives. Their data tells the story for them—accurate or not. Their opportunities become tied to systems they cannot see, challenge, or correct. The entity that owns the data gains the authority to shape the trajectory of a person's life.

The most alarming aspect is that this ownership is largely invisible. People give away data through consent forms they do not read, settings they do not understand, and systems they cannot opt out of without sacrificing participation in modern life. The transfer of ownership is hidden behind convenience. A person receives a free service, but the cost is their behavioral essence. This trade is rarely recognized, yet it has profound implications. When someone entrusts their data to another entity, they surrender control over how that data will be used, interpreted, or weaponized in the future.

Moreover, data ownership scales exponentially. Companies do not own one person's data—they own data from millions or billions of individuals. This creates power structures unmatched in human history. Never before has it been possible for a small number of entities to understand—and influence—the behavior of entire populations with such precision. The ownership of data becomes ownership of collective human behavior.

Ultimately, when data becomes property, people become projections—profiles molded according to what serves the goals of data owners. Autonomy becomes compromised not because individuals are forced to behave a certain way, but because systems nudge, predict, and shape them using the intimate details of their own lives. Ownership of data becomes ownership of influence, identity, opportunity, and direction.

In this sense, losing control of personal data is not merely a privacy issue—it is a sovereignty issue. It marks the moment when the boundaries of the self extend beyond the body and into the digital realm, and yet those boundaries are controlled by someone else. When others own the most intimate map of who you are, they possess a form of ownership over *you*.

THE QUIET TRANSFER OF POWER

Power rarely shifts in dramatic, visible strokes. In modern society, the most consequential transfers of power happen quietly—without announcement, confrontation, or public debate. This subtle shift is particularly evident in the digital age, where individuals gradually relinquish control not through force, but through convenience. The systems people depend on for communication, navigation, entertainment, shopping, and even self-expression have become the very systems that reshape autonomy. The transfer of power is not seized; it is surrendered in exchange for speed, simplicity, personalization, and connection.

The quietness of this transfer is what makes it so effective. People rarely notice when influence becomes dependence. When technology simplifies a task, it becomes indispensable. When a platform provides access, it becomes authoritative. When artificial intelligence predicts behavior accurately, it becomes trusted. Every small surrender of control feels harmless—until the cumulative loss becomes too large to reverse. By the time individuals realize the extent of this transfer, the power imbalance is already established.

This shift begins with reliance. As daily life becomes more digitized, people rely more heavily on platforms that store data, interpret behavior, and automate decisions. Each reliance reduces personal agency. For example, relying on apps for recommendations reduces the instinct to explore independently. Relying on automated filters reduces the practice of critical thinking. Relying on algorithms to curate information reduces the habit of seeking diverse perspectives. Small conveniences gradually replace once-essential human skills. This dependency forms the psychological groundwork for transferring power.

Another mechanism of quiet power transfer is opacity. Users do not understand how algorithms work, what data they collect, how decisions are made, or who ultimately benefits. This lack of transparency creates a one-sided power structure: individuals share everything, while systems reveal nothing. People operate within environments shaped by entities they cannot question, influence, or fully comprehend. Yet because the systems produce desirable outcomes—relevant content, personalized suggestions, efficient tools—individuals accept the opacity as part of the experience.

The transfer of power also thrives on emotional subtlety. Technologies are designed to feel helpful, comforting, or intuitive. Interfaces are smooth, notifications are gentle, and recommendations appear thoughtful. The digital environment is crafted to evoke trust. But beneath the aesthetics lies the reality that these systems are engineered to predict, influence, and retain attention. The warm surface hides the mechanisms of control. People surrender power not because they believe they must, but because they feel the system understands them.

Over time, this feeling transitions into a form of submission. When users repeatedly accept algorithmic decisions as superior—more accurate, faster, less effortful—they begin to defer their own judgment. They outsource decisions to machines, trusting them more than their own instincts. This deference transfers authority from the self to the system. The individual remains present, but their agency weakens. The system becomes the guide, the arbiter, the silent decision-maker.

As this dynamic strengthens, power shifts from individuals to the institutions that operate these systems. Companies gain the ability to shape behavior at scale, influence collective opinion, and direct attention with unprecedented precision. Governments increasingly rely on technology companies for surveillance, data processing, and public decision-making. The lines between corporate power and societal power blur. The authority once distributed across communities, institutions, and individuals becomes concentrated in technological infrastructures controlled by a few.

One of the most profound effects of this transfer is the redefinition of consent. People believe they are consenting when they click "agree," but true consent requires understanding—and most users do not fully understand what they are surrendering. Consent becomes symbolic, not substantive. The power dynamic shifts so gradually that individuals do not perceive the imbalance growing around them.

Another consequence is the diminishing ability to resist. As society's infrastructure becomes increasingly digital, opting out becomes impractical or impossible. You cannot function without digital identification, digital communication, digital banking, or digital navigation. Dependency erodes alternatives. When the only functioning systems are those that monitor, categorize, and influence individuals, power has been fully transferred—even if no one declared it so.

The final stage of this quiet transfer occurs when individuals internalize the system's influence. They adjust their behavior not because they are forced to, but because they anticipate how the system might interpret it. Self-censorship, conformity, and compliance become normalized. People obey not because they fear punishment, but because the digital environment subtly teaches them what is acceptable. This psychological compliance marks the moment when power has fully shifted.

The transfer of power in the digital age is not a revolution—it is a redesign. And unless individuals recognize how silently it happens, autonomy will continue to erode while convenience tells them nothing has changed.

Chapter 11

Who Controls the System Controls the Future

In a world driven by data, algorithms, and interconnected digital infrastructures, power no longer belongs solely to those who hold political office, economic wealth, or military strength. Instead, a new kind of authority has emerged—one built on the ability to design, operate, and manipulate the systems that define how people communicate, consume information, form beliefs, and make decisions. The entities that control digital systems do not merely influence the present; they shape the trajectory of the future.

This chapter explores a profound truth: whoever controls the system controls the narrative, and whoever controls the narrative controls the direction in which societies evolve. Corporations, governments, and technological institutions now function as invisible architects of human experience. Their influence is subtle yet immense, woven into algorithms, platforms, policies, and digital environments that people use every day without questioning who built them—or why.

Corporate entities exert power not through overt force, but through infrastructure. Platforms determine what information rises to visibility and what fades into obscurity. Search engines decide which knowledge becomes accessible. Social media companies shape cultural climates. Tech giants influence global behavior more deeply than many political systems ever could.

Meanwhile, governments recognize the strategic importance of digital control. Through regulation, surveillance, partnerships, and digital governance, states increasingly compete for influence over technological futures. Their involvement blurs the boundary between public authority and private power, creating a hybrid ecosystem where both cooperate—and sometimes clash—to shape the digital landscape.

At the heart of this transformation lies a critical question: who decides what becomes reality in a world mediated by screens, data, and algorithms?

This chapter examines the dynamics of power in the digital age, revealing how unseen forces shape the future long before individuals realize decisions have already been made.

CORPORATE INFLUENCE AND INVISIBLE GOVERNANCE

In the digital age, corporations no longer function merely as service providers or technology innovators—they have evolved into governing forces that shape human behavior, societal norms, and global power structures. Unlike traditional governments, whose authority is explicit, public, and constrained by laws, corporate influence operates through infrastructure, design, data ownership, and algorithmic control. This influence is subtle yet vast, forming an invisible governance system that directs how billions of people interact with the world.

Corporate power begins with control over digital environments. Platforms such as search engines, social networks, cloud services, app ecosystems, and communication tools form the backbone of modern life. These platforms determine what information is accessible, what content is amplified, what ideas spread, and what forms of expression are allowed. They decide which products people see, which news stories gain traction, and which voices become influential. By shaping the architecture of digital experience, corporations shape the architecture of thought.

This governance is invisible because it is embedded in functionality. The user does not see governance—they see convenience. When a platform decides which posts appear first, it does not announce, "We are governing your reality." It simply presents a feed that feels natural. When a search engine filters results, it does not declare, "We are selecting your knowledge." It simply displays what feels relevant. People accept these decisions because they appear neutral, logical, and personalized. But behind every design choice is a business objective, a strategic calculation, or an algorithmic preference.

One of the most significant forms of invisible corporate governance comes from algorithmic curation. Algorithms decide what individuals engage with, but individuals often believe they are choosing freely. In truth, the system presents options in a calculated order, shaping what appears important or desirable. This subtle leadership influences consumer habits, political attitudes, cultural trends, and emotional states. The corporation becomes a silent editor of human experience, with the power to guide behavior without overt instruction.

Another area of corporate governance involves the ownership and interpretation of data. Corporations collect vast amounts of information—behaviors, emotions, habits, vulnerabilities, and preferences. This data becomes the raw material for prediction and manipulation. The more data a corporation possesses, the more accurately it can forecast user actions and shape future ones. Data becomes both insight and leverage. Ownership of data becomes ownership of influence.

Corporations also govern through infrastructure dependency. Modern life is built on systems controlled by a handful of companies: servers, operating systems, payment gateways, digital identity systems, cybersecurity frameworks, cloud storage, and communication networks. If these systems fail—or if access is restricted—society itself is disrupted. This gives corporations a form of power traditionally held by governments: the ability to determine who can participate in the modern economy, who can speak, and who can be silenced. Corporate governance becomes a structural truth, not a metaphor.

Invisible governance is reinforced through design. Interface design influences behavior as effectively as laws or regulations. The placement of a button, the color of a notification, the timing of an alert—all are engineered to guide actions. Users believe they are interacting with tools, but the tools are interacting with them. Every tap, swipe, and scroll is shaped by corporate intention. This creates a feedback loop: the design shapes behavior, behavior produces data, data informs new design, and the cycle strengthens corporate influence.

Economic power also expands corporate governance. Technology companies hold financial resources that rival or exceed those of many nations. Their ability to acquire competitors, shape markets, and influence global supply chains gives them strategic leverage over entire sectors of society. They can set industry standards, negotiate with governments, and determine innovation trajectories. When corporations wield this level of power, governance becomes distributed—but not democratically.

Corporate governance extends into the political sphere as well. Through lobbying, partnerships, public influence, and control of communication platforms, corporations shape political discourse. They determine which voices are amplified, which movements gain visibility, and which narratives dominate public attention. Elections, social conflicts, cultural movements, and ideological shifts are all influenced, directly or indirectly, by corporate algorithms and policies.

What makes this governance particularly potent is the lack of accountability. Corporations are not elected, their processes are not transparent, and their priorities are shaped by profit rather than public welfare. Yet they control systems that define reality for billions. Governance without accountability becomes a quiet form of domination—not through oppression, but through design and dependence.

Invisible corporate governance is the new power structure of the digital age: subtle, pervasive, and immensely influential. And unless society recognizes the scale of this influence, corporate control over digital life will continue to expand, reshaping the human future without consent or awareness.

GOVERNMENTS, REGULATIONS, AND DIGITAL POWER

Governments have always exercised power through laws, enforcement, and institutional authority. But in the digital age, the nature of governance has changed. Power no longer lies solely in physical borders or traditional institutions; it now resides in data infrastructures, technological standards, and the ability to interpret and control information at scale. Governments around the world increasingly recognize that digital systems shape societies as profoundly as political systems do—and in some cases, even more. As a result, a new form of governance has emerged, one that blends state authority with technological control.

Governments exert digital power in several ways. The most obvious is surveillance. States monitor communications, track online behavior, collect metadata, analyze patterns, and store vast archives of digital activity. These tools, justified in the name of public safety or national security, give governments an unprecedented ability to observe and predict citizen behavior. Surveillance no longer requires physical informants or manual tracking; algorithms and automated systems perform the function at scale. A monitored population becomes a compliant one, not because force is applied, but because fear and uncertainty shape behavior.

Regulations also serve as a major avenue of digital power. Governments influence how corporations operate, how data is collected, how platforms moderate content, and how information flows across borders. These policies determine what citizens are exposed to, what is suppressed, and what narratives gain legitimacy. When states regulate digital space, they shape the cognitive environment of society. Regulations become ideological tools, shaping collective perception in the same way that traditional propaganda once did.

However, governments face a paradox: they depend on private technology companies to execute digital governance. Platforms control the infrastructure—servers, software, communication networks, algorithms, and data centers—that governments cannot easily replace. As a result, states often collaborate with corporations, requesting access to data, demanding compliance with policies, or pressuring companies to moderate content in ways that serve political goals. This merging of corporate capability and governmental authority creates a hybrid power structure, where neither entity operates independently. Citizens become subject to a system in which transparency is limited and accountability is diffuse.

Another layer of digital power involves shaping national identity and public discourse. Governments influence what information circulates through media channels, state-sponsored narratives, and algorithmic partnerships. In some nations, information control becomes explicit, with censorship, firewalls, and content filtering defining the digital environment. In others, the influence is more subtle—shaped through regulatory pressure, political partnerships, and the strategic use of digital platforms to amplify preferred messages. In both cases, the result is the same: digital reality becomes a tool of governance.

Digital power also extends to economic influence. Governments that control technological development—artificial intelligence, semiconductors, cybersecurity systems, telecommunications—gain strategic leverage domestically and globally. Nations compete for dominance in these areas because they understand that technological leadership translates into political dominance. Countries without control over digital infrastructure risk becoming dependent on foreign technologies, losing sovereignty not to other governments but to global corporations and technological ecosystems.

Cybersecurity further amplifies governmental power. The ability to defend against cyberattacks, or to launch them, becomes a form of geopolitical influence. Cyber warfare, espionage, infrastructure attacks, and digital sabotage redefine the battlefield. Digital threats create fear, and fear consolidates power. States justify deeper surveillance, expanded authority, and emergency regulations through the narrative of cyber risk. While some of these measures may be necessary, others quietly shift the balance of freedom and control.

One of the most significant expressions of governmental digital power is algorithmic governance—the use of algorithmic systems to make decisions previously made by humans. Governments deploy algorithms for policing, welfare distribution, immigration decisions, risk assessment, and resource allocation. These automated systems appear neutral but often inherit biases, limitations, and assumptions that shape real-world outcomes. When algorithms determine access to essential services, digital power becomes a form of structural authority—one that is difficult for citizens to challenge or question.

The rise of digital power challenges traditional notions of democracy. When information is filtered through algorithms shaped by political interests, public opinion becomes manipulable. When surveillance is pervasive, dissent becomes risky. When regulations shape digital expression, the line between governance and control blurs. Citizens may still vote, but the environment in which they form opinions has already been influenced by forces they do not see.

The future of governance will depend on how societies balance innovation with accountability, security with freedom, and digital power with democratic transparency. If governments use digital tools to empower citizens, technology becomes a force for progress. If they use them to monitor, manipulate, or suppress, technology becomes a modern instrument of authoritarianism.

Digital power is not inherently good or evil—it is transformative. And whoever controls it will shape not just policy, but the very experience of reality for generations to come.

DIGITAL REALITY AS A CONTROLLED NARRATIVE

Digital reality is not a passive reflection of the world—it is an engineered environment shaped by algorithms, corporate priorities, political pressures, and psychological design principles. The information people encounter online forms a curated version of reality, one that feels natural and neutral but is structured intentionally. This curated environment influences what people know, what they believe, how they interpret events, and ultimately how they behave. In the digital age, narratives are not merely communicated—they are constructed, optimized, and continuously reinforced by the systems that deliver them. Whoever controls these systems controls the narrative, and whoever controls the narrative controls the future.

The most fundamental aspect of narrative control is the filtering of information. No platform shows everything; each system selects what is "relevant," "engaging," or "important" based on algorithmic criteria. These criteria are invisible to users, yet they determine which stories rise to the surface and which disappear. A person's perception of the world becomes shaped by what the platform chooses to highlight. This filtering shapes the emotional and intellectual climate of society, influencing public opinion without citizens realizing the extent of the manipulation.

Algorithms also amplify certain narratives through repetition. Repeated exposure gives ideas power by making them feel familiar and therefore true. Even neutral algorithms inadvertently strengthen narratives that encourage engagement—anger, fear, outrage, identity conflict—because these emotions keep users active. Platforms do not need to intentionally promote certain perspectives; their design naturally elevates content that captures attention. This unintentional bias creates a distorted reality in which emotionally charged narratives dominate, and quieter, more nuanced perspectives are buried.

Another layer of narrative control involves personalization. People do not experience the same digital world; they experience individualized streams of information tailored to their preferences, behaviors, and psychological tendencies. Personalization reinforces existing beliefs, forming echo chambers where individuals rarely encounter opposing viewpoints. These personalized realities create fragmented societies, each living inside its own narrative bubble. People begin to believe their version of the world is the only version. This makes manipulation easier, conflict more common, and consensus more difficult.

The structure of digital platforms also shapes how narratives are experienced. Short-form content, infinite scroll, visual emphasis, and rapid consumption all encourage shallow engagement. Complex issues become simplified, stripped of context, and transformed into digestible fragments that fit the platform's format. This leads to a reduction in critical thinking. People become reactive rather than reflective—guided not by deep understanding but by emotional triggers engineered to maximize engagement. When depth is replaced by speed, narratives become tools of influence rather than pathways to truth.

Digital reality also allows powerful institutions—corporations, governments, influencers, and interest groups—to strategically craft narratives. Through targeted advertising, coordinated messaging, content shaping, and emotional manipulation, they can steer public perception with unprecedented precision. In earlier eras, propaganda required mass broadcasting; today it can be personalized down to the individual psychological profile. The ability to shape beliefs at scale without the subject even realizing they were targeted transforms narrative into a form of silent governance.

Influencers and creators also play a role in constructing digital reality. Many become extensions of corporate or political agendas, consciously or unconsciously shaping the values and attitudes of their audiences. Their content appears authentic, personal, and organic, making it more persuasive than traditional media. Yet behind their influence lies a complex web of algorithms that determine visibility, reach, and public standing. Those who align with platform incentives rise; those who challenge dominant structures risk invisibility.

The collapse of authoritative truth is another consequence of narrative control. When digital platforms elevate all voices equally, expertise becomes diluted. Facts and misinformation coexist in the same space, distinguished not by accuracy but by engagement. This creates a chaotic informational landscape where people struggle to determine what is real. In such an environment, narratives gain power not because they are true, but because they are repeated, emotionally appealing, or aligned with algorithmic incentives.

Digital reality is not inherently manipulative; it is malleable. But this malleability becomes dangerous when shaped primarily by entities seeking profit, influence, or control. Without transparency, accountability, or public awareness, digital narratives become instruments of subtle domination. People believe they are forming independent opinions, unaware that their informational environment has been engineered around them.

Ultimately, digital reality functions as a controlled narrative because it shapes the boundaries of what people see, what they think about, and what they imagine possible. The future will be determined not only by technological innovation but by who controls the stories, symbols, and structures that define reality itself.

Chapter 12

Reclaiming Human Choice in a Designed World

In a world shaped by algorithms, surveillance systems, psychological design, and invisible power structures, human choice no longer emerges from a neutral environment. It is influenced, guided, nudged, predicted, and at times quietly controlled by the digital ecosystems people inhabit daily. Yet even within this engineered landscape, the possibility of reclaiming authentic choice remains. The first step is recognizing that free will no longer exists by default—it must be intentionally cultivated.

This chapter explores the pathways toward restoring autonomy in a world that subtly redesigns behavior. Awareness becomes the initial act of resistance. When individuals understand how digital systems shape perception and influence decisions, the invisible becomes visible. What once operated unconsciously can be questioned, interrupted, and re-evaluated. Awareness transforms people from passive participants into active observers of their own mental environment.

But awareness alone is not enough. To reclaim human choice, individuals must build boundaries—psychological, technological, and behavioral. These boundaries protect the mind from manipulation, create space for reflection, and re-establish a sense of agency. Setting limits around technology use, controlling informational exposure, and developing habits of intentional thinking become modern forms of self-defense.

Finally, reclaiming choice requires a deeper shift: living consciously. This means making decisions based on values rather than impulses, purpose rather than convenience, and self-direction rather than algorithmic suggestion. Conscious living transforms everyday actions into expressions of autonomy, re-centering the individual in a world that constantly pulls them outward.

This chapter does not argue for abandoning technology; it argues for mastering it. Freedom in the digital age does not arise from escaping systems, but from understanding them well enough to choose deliberately within them.

AWARENESS AS THE FIRST RESISTANCE

Awareness is the first and most essential act of resistance in a world where influence is invisible, persistent, and woven into the digital environments people depend on. Most forms of manipulation succeed not because they are powerful, but because they are unnoticed. When individuals become aware of how algorithms shape behavior, how platforms engineer attention, and how systems influence thought, the unconscious grip of these forces weakens. Awareness restores the ability to pause, question, and choose. It is the moment when automatic responses begin to dissolve and intentional living begins to take root.

To understand awareness as resistance, one must first recognize that modern manipulation is subtle. It does not announce itself. It does not look like control. It feels like convenience, relevance, personalization, and efficiency. What appears helpful may actually be directive. What feels natural may have been engineered. Awareness reveals these layers. It helps individuals distinguish between genuine desire and algorithmically shaped impulses, between personal thought and curated information, between self-driven behavior and environmental conditioning.

Awareness also interrupts the automaticity that technology depends on. Many digital systems are designed to operate beneath conscious attention. Infinite scrolls, dopamine-triggering notifications, habit loops, and predictive suggestions bypass deep thinking. They rely on reflex rather than reflection. But when a person becomes aware of these mechanisms—when they notice the pull, the pattern, the trigger—the spell breaks. Awareness creates mental distance between stimulus and response. In that distance, freedom appears.

Another function of awareness is recognizing emotional vulnerability. Algorithms often target emotions—loneliness, boredom, insecurity, anger, curiosity—because emotional states increase engagement. When individuals understand how their emotions are exploited, they can begin to protect them. They can ask: Why am I seeing this now? What mood am I in? How is this content trying to influence me? Awareness transforms emotion from a point of vulnerability into a point of strength.

Awareness also exposes the architecture of digital environments. Most people never question why certain buttons are placed where they are, why certain colors feel urgent, why suggestions appear at specific moments, or why content feels perfectly tailored. They assume the design is neutral. In reality, design is persuasion. Awareness reveals that every interface is a psychological landscape crafted to guide behavior. Once individuals see this, they stop being passive participants and start becoming analysts of their environment.

Another form of awareness involves understanding personal data trails. Many people do not know what information is collected about them, how it is interpreted, or how it shapes future experiences. When individuals realize that their data forms a digital replica of themselves—one used to predict and influence behavior—they become more mindful of what they share, how they engage, and what permissions they grant. Awareness encourages digital caution, not out of fear but out of informed self-protection.

Awareness also dismantles the illusion of neutrality. Digital platforms present themselves as objective tools, but they are built on business models, political pressures, and psychological strategies. Once individuals understand this, they stop assuming that what they see is simply "what's happening." They begin to question whose interests are being served, whose voices are amplified, and whose perspectives are hidden. Awareness restores critical thinking by encouraging skepticism in an environment engineered for trust.

One of the most empowering aspects of awareness is that it spreads. When one person becomes aware, they influence others. Conversations shift. Families discuss digital habits. Friends share insights about manipulation. Communities become more vigilant. Awareness creates collective resilience. It transforms isolated individuals into informed participants capable of resisting systemic influence.

But awareness must be continuous. Manipulation evolves. Algorithms adapt. Platforms redesign themselves. New technologies emerge. Awareness is not a single realization—it is an ongoing practice of observing oneself and one's environment. It requires noticing patterns, reflecting on impulses, questioning assumptions, and remaining alert to subtle shifts in influence. Awareness becomes a mental posture: calm, observant, and intentional.

Most importantly, awareness reconnects individuals with their internal world. It redirects attention inward, helping people rediscover their genuine desires, values, and intuitions. In a world of external noise, awareness becomes a sanctuary—a space where the self is no longer drowned out by algorithms. Awareness does not eliminate manipulation, but it reduces its power. It does not create perfect freedom, but it creates the possibility of freedom. It allows individuals to reclaim the first and most fundamental human right: the right to one's own mind.

Awareness is resistance because it interrupts the forces that operate without permission. It is the beginning of autonomy, the first crack in the illusion of control, and the foundation upon which all other forms of freedom are built.

BUILDING BOUNDARIES AGAINST MANIPULATION

Building boundaries in a world engineered to bypass them is one of the most important acts of self-preservation. Manipulation thrives when individuals have no defenses, no structure around their attention, no clarity about their vulnerabilities, and no limits on the systems allowed to shape their thoughts. Boundaries create separation between the self and the environment. They restore agency by giving individuals control over what reaches their mind, how much influence it gains, and what psychological space it occupies. Boundaries are not barriers to progress—they are guardrails for autonomy.

The first kind of boundary is informational. In a digital world overflowing with curated feeds, targeted content, and algorithmic suggestions, people must regulate the information they consume. Not every piece of digital input deserves access to the mind. Informational boundaries include choosing reliable sources, limiting exposure to emotionally manipulative content, and questioning the intention behind what appears on a screen. Individuals must develop the discipline to pause before accepting new information and ask, "Why am I being shown this? Who benefits from my attention?" By filtering input consciously, people prevent manipulation from entering through the front door of awareness.

Another critical boundary is technological. Devices are designed to demand attention, interrupt thought, and create habit loops. Without boundaries, technology becomes a constant intruder in the mental space. Setting limits—turning off non-essential notifications, creating phone-free spaces, scheduling intentional digital breaks—protects the mind from the endless pull of engineered engagement. These boundaries weaken the power of persuasive design by reducing the number of opportunities systems have to trigger impulsive responses. Technology should serve the individual, not occupy them.

Psychological boundaries are equally important. Manipulation often exploits emotional vulnerabilities such as loneliness, fear, boredom, insecurity, or validation-seeking. By becoming aware of these states and treating them with care, individuals reduce the influence of external manipulation. Emotional awareness allows people to recognize when they are most susceptible to persuasion. Instead of turning to digital platforms for distraction or comfort, individuals can develop healthier grounding practices—reflection, conversation, physical activity, mindfulness, or creative expression. Strengthening emotional resilience creates a barrier against the systems that feed on emotional instability.

Another form of boundary involves reclaiming attention. Attention is the currency of the digital world, and whoever controls it controls behavior. Setting boundaries around attention—such as focusing on one task at a time, avoiding multitasking, and reclaiming moments of silence—reduces the influence of digital stimuli. Attention boundaries also include intentional consumption: choosing when to engage rather than reacting instantly to digital prompts. A person who controls their attention controls their choices; a person who loses their attention loses their autonomy.

Cognitive boundaries are also essential. These boundaries involve maintaining mental independence by questioning assumptions, challenging easy narratives, and thinking beyond algorithmically curated perspectives. Cognitive boundaries protect individuals from being shaped by the dominant digital narratives of a platform. They encourage deeper reasoning, slower thought, and personal interpretation. Instead of absorbing the world through the lens of algorithmic relevance, individuals can reclaim the ability to create their own worldviews.

Behavioral boundaries include establishing routines that support autonomy—such as designated times for digital use, limits on social media, and structured moments of self-reflection. These routines reduce the subconscious influence of digital platforms by placing intentional behavior ahead of automatic habits. When individuals structure their time deliberately, they weaken the power of systems designed to consume their time without permission.

Social boundaries are also critical. Digital manipulation often operates through social comparison, validation-seeking, and the desire for acceptance. Limiting exposure to environments that trigger unhealthy comparison, setting boundaries around online interactions, and intentionally curating social circles protect the mind from social pressures amplified by digital systems. Social boundaries help individuals maintain a stable sense of identity—one not determined by the shifting tides of digital approval.

One of the most powerful boundaries is silence. Modern life leaves little room for quiet reflection. Systems of manipulation thrive on noise—constant input, endless distraction, emotional stimulation. Silence provides the mental stillness necessary to reflect, evaluate, and reset. Through silence, individuals reconnect with their values, intentions, and inner clarity. Silence becomes a shield against environments that seek to drown the self in external noise.

Ultimately, boundaries must be built not as rigid constraints but as living practices. The digital world evolves, and boundaries must evolve with it. Manipulation becomes weaker when met with awareness, intentionality, and self-control. Boundaries do not eliminate influence, but they prevent influence from becoming domination. They create the psychological and emotional space necessary for genuine choice.

In a world designed to infiltrate the mind, boundaries are acts of liberation. They do not isolate individuals from technology but empower them to use it on their own terms. Boundaries restore the separation between human intention and external design—allowing the self to breathe, think, and choose freely once again.

RESTORING FREEDOM THROUGH CONSCIOUS LIVING

Restoring freedom in a world engineered to shape behavior requires more than awareness and boundaries—it requires a transformation in how individuals live, choose, and relate to their own minds. Conscious living is the antidote to unconscious influence. It is the practice of reclaiming agency by making deliberate choices instead of automated ones, by acting from values rather than impulses, and by reorienting life toward intention rather than digital design. In a world that constantly pulls attention outward, conscious living brings it back inward where freedom originates.

Conscious living begins with intentionality. Many people move through their days guided not by inner purpose but by external cues—notifications, trending topics, algorithmic suggestions, emotional triggers, and habitual routines. Intentional living requires pausing long enough to ask: "Is this what I choose, or what I'm being pulled toward?" This question breaks the automatic loop that technology depends on. It inserts human intention into the sequence of stimulus and response. The more often individuals return to intention, the weaker external manipulation becomes.

Another part of conscious living is reconnecting with one's inner world. Modern life, especially in digital spaces, prioritizes reaction over reflection. People are encouraged to respond instantly, consume rapidly, and move continuously from one stimulus to the next. But conscious living creates space for reflection—moments to think deeply, evaluate meaning, and reconnect with personal values. Reflection restores clarity. It helps individuals distinguish between desires shaped by digital conditioning and those that reflect authentic aspiration. Without reflection, autonomy becomes fragmented; with reflection, it begins to reassemble.

Conscious living also requires slowing down. Speed is one of the digital world's greatest weapons. When people rush, they do not notice manipulation. They do not pause to question. They do not give themselves time to choose. Slowing down—whether in thought, speech, decisions, or consumption—restores control. It allows space for discernment, for recognizing emotional triggers, and for resisting impulsive behavior. Freedom does not live in speed; it lives in intentional pace.

A key element of conscious living is developing mindful awareness of emotional states. Manipulation thrives on emotional instability—fear, loneliness, boredom, insecurity, envy, outrage. Conscious living encourages people to observe their emotions rather than be pulled by them. When individuals know how they feel and why they feel it, they can prevent external systems from exploiting those emotions. Instead of reacting automatically to emotional triggers, they respond with clarity and self-direction. Emotional mindfulness becomes a shield against psychological influence.

Values act as another compass for conscious living. In a world that constantly tells people what to want, what to believe, and what to become, values serve as an internal anchor. Conscious living requires articulating personal values and using them to guide decisions. When choices align with values, they become stable and resilient. When choices are shaped by external pressure, they become fragile and reactive. Living from values restores autonomy because decisions become expressions of identity rather than responses to manipulation.

Conscious living also includes mastering attention. Attention is the gateway through which influence enters. By choosing where to place attention—and what to withdraw it from—individuals reclaim control over their mental environment. This might involve choosing fewer inputs, consuming content slowly, or practicing deep focus. The more intentional a person becomes about their attention, the less likely they are to be driven by algorithmic design. Attention mastery reclaims the mind from digital environments that seek to fragment it.

Additionally, conscious living requires cultivating presence. Presence counters the dissociation created by constant digital engagement. It encourages individuals to experience life directly rather than through screens, to notice their surroundings, and to reconnect with their bodies, senses, and relationships. Presence strengthens autonomy because it brings awareness back to the immediate moment, where real choice exists. When individuals are fully present, they are less susceptible to psychological drift.

Conscious living also involves designing supportive environments. People shape their environments, and their environments shape them. By creating physical and digital spaces that support clarity—spaces free from constant alerts, unnecessary noise, or emotional manipulation—individuals nurture their autonomy. Conscious living is not only about resisting influence but about cultivating environments that make authentic living easier.

Ultimately, conscious living restores freedom by shifting the source of decisions from external design to internal intention. It does not reject technology but reshapes the relationship with it. Instead of being pulled, individuals choose. Instead of absorbing, they filter. Instead of reacting, they reflect. Conscious living is not a single act—it is a continual practice of returning to oneself.

In a world designed to shape behavior, conscious living becomes a revolutionary act. It reclaims the mind as sovereign, restores dignity to human choice, and reawakens the deep, inherent freedom that external systems can influence but never fully extinguish.

www.ingramcontent.com/pod-product-compliance
Lightning Source LLC
Chambersburg PA
CBHW041300210926
51847CB00009B/87
9789699797378